Physics Workbook Volume 1

Revised Summer 2022

The solutions are available <u>for free</u> as a pdf at http://www.robjorstad.com/Phys161/161Workbook.htm.

Use a laptop, tablet, or phone to access the solutions at my website above.

Find the chapter you want then open that link on your device.

The questions (this book) and the answers (on your device) ready to go at the same time!

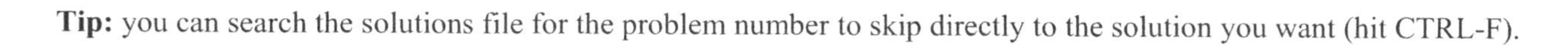

Tip: you can search the solutions file for the problem number to skip directly to the solution you want (hit CTRL-F).

SIG FIGS, SCIENTIFIC NOTATION, UNITS, ETC

Why pay attention to sig figs?

Every time you write down a number, the number of sig figs implies the precision of that number. A number with 2 sigs figs has *about* 10% uncertainty while a number with 3 sig figs has *about* 1% uncertainty. Four sig figs imply approximately 0.1% uncertainty; five sig figs imply about 0.01% uncertainty…extremely precise.

- Writing numbers with too few sig figs causes errors in subsequent calculations.
- Writing numbers with too many sig figs implies a false sense of precision of the answer.

The best technique is to write one extra sig fig but indicate the uncertain digit. This allows you to avoid intermediate rounding errors while still properly communicating to others the true level of precision in your calculations.

To determine the number of sig figs in a number:

Rule 1: All non-zero digits are significant.
Rule 2: Zeros between other significant figures (bounded zeros) are significant.
Rule 3: Leading zeros are to the *left* of all non-zero numbers. They are never significant.
Rule 4: Trailing zeros are *right* of all non-zero numbers. If all trailing zeros are to the *left* of the decimal, none are significant. **IF** at least *one* trailing zero appears to the *right* of the decimal point, then ALL trailing zeros are significant!

Example 1: Determine the number of sig figs in 40300.
Notice, in this case, the trailing zeros are NOT significant since none of them lie to the right of the decimal.

Example 2: Determine the number of sig figs in 0.000003600 in scientific notation.
Notice, in this case, the trailing zeros ARE significant since they lie to the right of the decimal. Note: *leading* zeros are never significant.

Example 3: Determine the number of sig figs in 340.0.
Notice, in this case, the trailing zeros ARE significant. The trailing zero to the *right* of the decimal is significant. The trailing zero to the *left* of the decimal is between two significant figures and thus becomes significant! Trailing zeros are all significant if any one of them lies to the right of the decimal!

1.1 Identify the number of sig figs in each number using the methods described in class.

a	123	d	0.03	g	1.00
b	10.0	e	1.03	h	$10\underline{0}0$ (the underbar indicates that zero *is* significant)
c	103	f	0.030	i	100

1.2 Write the following numbers in scientific notation. If you have any trouble, try reading the next page.

a	203	b	200	c	0.03030

Scientific Notation

A number written in scientific notation is expressed with a number between 1 and 10 multiplied by 10 to some power. Essentially, one must write the number so there is only one non-zero number to the left of the decimal.

Example 1: Write 40300 in *scientific* notation.

$$\underleftarrow{4030}0 = 4.03 \times 10^4$$

Example 2: Write 0.000003600 in *scientific* notation.

$$0.\underrightarrow{000003}600 = 3.600 \times 10^{-6}$$

Example 3: Write 340.0 in *scientific* notation.

$$3\underleftarrow{40}.0 = 3.400 \times 10^2$$

1.3 Write the following numbers in *scientific* notation.

Number	# of sig figs	Scientific Notation
0.000354		
80.5		
80.0		
1234		
12		
0.000000030		
0.45600		
6700		
860		
860.0		

1.4 Write the following numbers *without scientific* notation.

Tip: use the EE &mode buttons on your scientific calculator to check your work…ask your instructor.

Scientific Notation	# of sig figs	Number
1.23×10^{-4}		
3.00×10^{0}		
1.000×10^{-1}		
2.0×10^{5}		

Engineering Notation

Engineering notation writes numbers like this

$$\# = a \times 10^{n}$$

Here a is a number between 1 and 1000 and n is some multiple of three. Engineering notation, at first glance, looks nearly identical to scientific notation. In fact, sometimes it is exactly the same.

The difference between *scientific* & *engineering* notation:

When moving the decimal in *engr* notation, do it in groups *of three* and end with a number between 1 and 1000.

Example 1: Write 40300 in *engineering* notation. Move decimal over 3.

$$40300 = 40.3 \times 10^{3}$$

Example 2: Write 0.000003600 in *engineering* notation. Move decimal over $3 \times 2 = 6$.

$$0.000003600 = 3.600 \times 10^{-6}$$

Example 3: Write 4.01×10^{-2} in *engineering* notation. First write in normal mode, then move over 3 places.

$$4.01 \times 10^{-2} = 0.0401$$
$$0.0401 = 40.1 \times 10^{-3}$$

Example 4: Write 1.230×10^{8} in *engineering* notation. Move decimal over $3 \times 2 = 6$.

$$1.230 \times 10^{8} = 123\underline{0}00000$$
$$123\underline{0}00000 = 123.0 \times 10^{6}$$

Example 5: Write 3.400×10^{2} in *engineering* notation. First write in normal mode; then notice you are already done!

$$3.400 \times 10^{2} = 340.0$$
$$340.0 = 340.0 \times 10^{0}$$
$$no\ need\ to\ shift!$$

1.5 Write the following numbers in engineering notation.

Tip: use the EE &mode buttons on your scientific calculator to check your work…ask your instructor.

Number	Engineering Notation
0.000354	
80500	
0.00000003	
1234.0	
0.45600	
860.0	

Number	Engineering Notation
12	
9.09×10^{7}	
6.20×10^{-11}	
1.23×10^{-3}	
2.46×10^{4}	

Why bother with a prefix?
I prepared three data tables below. The numbers in each table are identical. The first uses scientific notation, the second uses engineering notation, the third uses engineering notation and an appropriate prefix. I think you will agree the third version just looks better.

x (m)
1.00E+03
2.00E+03
5.00E+03
1.000E+04
2.000E+04
5.000E+04
1.0000E+05
2.0000E+05
5.0000E+05

x (m)
1.00E+03
2.00E+03
5.00E+03
10.00E+03
20.00E+03
50.00E+03
100.00E+03
200.00E+03
500.00E+03

x (km)
1.00
2.00
5.00
10.00
20.00
50.00
100.00
200.00
500.00

In our society, scientists are expected to produce results AND communicate clearly to a wide range of audiences with non-technical backgrounds. The first two styles of results, while perfectly legitimate, will irritate non-technical audiences and make them stop listening to your message. Being able to write data in the third format makes you appear more professional. People will like you more. You will eventually get more money. Learn it! Oh yeah, your ability to choose an appropriate prefix will be tested on exams, too…

Prefix	**Abbreviation**	$\mathbf{10^?}$
Giga	G	10^9
Mega	M	10^6
kilo	k	10^3
centi	c	10^{-2}
milli	m	10^{-3}
micro	μ	10^{-6}
nano	n	10^{-9}
pico	p	10^{-12}
femto	f	10^{-15}

Example 1: $1.23\ \mathrm{nm} = 1.23 \times 10^{-9}\ \mathrm{m}$

Example 2: $0.456\ \mathrm{kcal} = 0.456 \times 10^{3}\ \mathrm{cal} = 456\ \mathrm{cal}$

Example 3: $0.0297\ \frac{\mathrm{m}}{\mathrm{s}} = 29.7 \times 10^{-3}\ \frac{\mathrm{m}}{\mathrm{s}} = 29.7\ \frac{\mathrm{mm}}{\mathrm{s}}$

Example 4: $1{,}860{,}000\ \text{phones} = 1.86 \times 10^{6}\ \text{phones} = 1.86\ \text{Mphones}$

Example 5: $70000\ \mathrm{m} = 70 \times 10^{3}\ \mathrm{m} = 70\ \mathrm{km}$

Example 6: $0.0350\ \mathrm{s} = 35.0 \times 10^{-3}\ \mathrm{s} = 35.0\ \mathrm{ms}$

Example 7: $0.00012\ \mathrm{A} = 120 \times 10^{-6}\ \mathrm{s} = 120\ \mu\mathrm{A}$

1.6 Write the following numbers in engineering notation with units! Also write them with the appropriate prefix. **Tip:** do these *without* a calculator first. Then type the number into calculator and change into engineering notation mode to check your work.

Number	Engineering Notation	Eng. Not. With Prefix
0.0434 m		
501000 V		
0.00000000020 C		
90000000 F		
0.77 g		
120000000000 Bq		
12 ft		
8.08×10^{8} J		
1.50×10^{-13} N		
1.23×10^{-8} bel		
2.46×10^{3} cd		

Why not always do this process with a calculator? When you write code, you might need to understand this process and you won't be able to use a calculator! As an example: suppose you obtain distance data (y) in mm but a calculation uses $g = 9.8\frac{\text{m}}{\text{s}^2}$. Perhaps the equation for time is $t = \sqrt{\frac{2y}{g}}$. In your computer code you would need to convert the prefix and write it as $t = \sqrt{\frac{2\left(\frac{y}{1000}\right)}{g}}$. **Don't worry about this paragraph now if it makes no sense.**

1.7 A slab of metal has length $L = 0.40$ m, width $w = 0.0808$ m, and thickness $t = 750 \times 10^{-7}$ m. Figure not to scale.

a) Write down L in *scientific* notation (correct sig figs and units).
b) Write down L in *engineering* notation with correct sig figs *and appropriate prefix*!
c) Write down w in *scientific* notation (correct sig figs and units).
d) Write down w in *engineering* notation with correct sig figs *and appropriate prefix*!
e) Write down t in *scientific* notation (correct sig figs and units).
f) Write down t in *engineering* notation with correct sig figs *and appropriate prefix*!

Math with sig figs

1. When multiplying or dividing, the crappiest *number* of sig figs is kept.
2. When adding or subtracting, the crappiest *column* of sig figs is kept.
   WARNING: Addition & subtraction can change the number of sig figs.
3. Unless otherwise specified, use *three* sig figs for everything.
   Exception: sometimes people default to *four* sig figs if the first digit of a number is 1.
4. In *scientific* notation, all numbers are always significant!
5. Keep at least one *extra* sig fig for all math work. Then, round you final results to the appropriate number of sig figs in the last step.

Avoiding intermediate rounding error

When doing computations, it is important to track sig figs as you go through the problem. The best method is to keep at least one extra sig fig on all numbers during your calculations and only round at the final step.

If you round your numbers *before* the final step, you introduce *intermediate rounding error*. By rounding too soon, your final answer can differ dramatically from the correct method.

I keep track of my sig figs using a small underline in each number. For example, the number $2.4\underline{3}6$ shows I have a 3 sig fig number (perhaps my calculator gave the extra digit after a computation).

Done Correctly Keep one extra sig fig, round after final answer	**Done Incorrectly** Round after each step
$x = \left[\frac{9.8}{6.2} - \frac{8.5}{7.5}\right]^2 - \frac{6.2}{8.5}$	$x = \left[\frac{9.8}{6.2} - \frac{8.5}{7.5}\right]^2 - \frac{6.2}{8.5}$
$x = [1.\underline{5}8 - (1.\underline{1}3)]^2 - 0.7\underline{2}9$	$x = [1.\underline{6} - (1.\underline{1})]^2 - 0.7\underline{3}$
$x = [0.\underline{4}5]^2 - 0.7\underline{2}9$	$x = [0.\underline{5}]^2 - 0.7\underline{3}$
$x = 0.\underline{2}03 - 0.7\underline{2}9$	$x = 0.\underline{3} - 0.7\underline{3}$
$x = -0.\underline{5}26$	$x = -0.\underline{4}$
$x = -0.\underline{5}$	Final answer differs by 20% from correct method!!!

1.8 Perform the following mathematical operations while keeping track of the correct number of sig figs. Write your final answer in scientific notation.

a) $\frac{120.0}{3} + 70 =$
b) $13 + 0.741 =$
c) $65.02 - 64.99 =$
d) $(12.0 - 9.99) \times (8.00 \times 10^6)$
e) $\frac{(13.10 - 13.00)^2}{(800+300)} =$

1.9 Explain the difference between 10E3 and 10^3. Try typing both numbers into your calculator now so you don't mess this up on an exam!

1.10 Answer the following subtraction problem with **correct significant figures** and **correct scientific notation**.

$$1.012 \times 10^4 \text{ km} - 9943.0 \text{ km} = ?$$

1.11 You are told $= 8769.8\underline{9}7$ N, $B = 8770.3\underline{2}4$ N, $C = 0.00083897$ m and $D = 0.000842\underline{2}4$ m.

a) Compute $B + A$. Answer in *engineering* notation *with appropriate prefix and correct sig figs.*
b) Compute $B - A$. Answer in *engineering* notation *with appropriate prefix and correct sig figs.*
c) Compute $D + C$. Answer in *engineering* notation *with appropriate prefix and correct sig figs.*
d) Compute $D - C$. Answer in *engineering* notation *with appropriate prefix and correct sig figs.*

1.12 A very rare physics unit is the mockingbird. $E = 4.97 \times 10^3$ mockingbird & $F = 20$ mockingbird. Compute $R = 346.5(F) - E$ with correct sig figs, engineering notation, & appropriate prefix.

1.13 Suppose you have a rectangular fence with sides of 4.0 m and 6.0 m. Determine the perimeter and the area of the fence. Answer with proper units, sig figs, and scientific notation.

1.14 You are given the data table shown at right and the formula below.

$$\Delta x = \frac{1}{2} a_x t^2 + v_{0x} t$$

$a_x \left(\frac{\text{m}}{\text{s}^2}\right)$	$v_{0x} \left(\frac{\text{m}}{\text{s}}\right)$	t (s)
-1.23×10^{-1}	5.85×10^{-3}	0.0987

Compute Δx while correctly keeping track of sig figs. Write your final answer in engineering notation with appropriate prefix. Correctly round (or indicate rounding column with the underbar).

- We will learn later that the ½ has infinite sig figs. For now, assume $\frac{1}{2} = 0.500\underline{0}$. By having more sig figs than any other number in the problem it will not affect the sig figs in the calculation.
- Carry along the units at each step and cancel out the units of seconds (s) as appropriate.

1.15 You are given the data table shown at right and the formula shown.

$$v_{fy}^2 = v_{iy}^2 + 2a_y \Delta y$$

$a_y \left(\frac{\text{m}}{\text{s}^2}\right)$	$v_{fy} \left(\frac{\text{m}}{\text{s}}\right)$	Δy (m)
-9.8	9.2×10^1	8.7×10^1

a) Solve algebraically for v_{iy}.
b) Next, plug in numbers form the data table to compute v_i. Write your final answer in engineering notation with appropriate prefix. Correctly round (or indicate rounding column with the underbar). Assume 2 has infinite sig figs.

Notice there are two questions at the bottom of this page.

STUFF YOU SHOULD KNOW ALREADY

Example 1: SOH CAH TOA as it applies to the triangle at right.

$$\sin\alpha = \frac{\text{opposite}}{\text{hypotenuse}} = \frac{B}{C}$$

Example 2:

$$\tan\beta = \frac{\text{opposite}}{\text{adjacent}} = \frac{A}{B}$$

Volume of a sphere: $V_{sphere} = \frac{4}{3}\pi R^3$
Surface Area of a Sphere: $A_{sphere} = 4\pi R^2$

Volume of a cylinder: $V_{cyl} = \pi R^2 H$

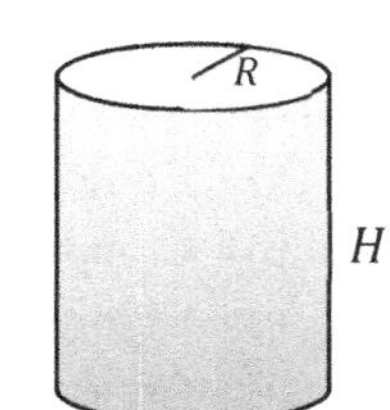

Volume of a rectangular box: $V_{box} = LWH$

Arbitrary shape volume: often (but not always) $V = (Area\ of\ base) \times (height)$

Area of rectangle: $A_{rect} = LW$

Area of circle: $A_{circle} = \pi R^2$
Circumference: $C = 2\pi R$

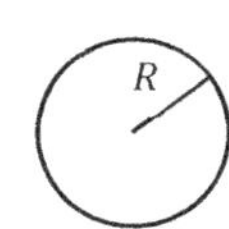

1.16 Use the above information to re-write the volume of a sphere in terms of diameter (D) instead of radius.

1.17 Use the above information to re-write the volume of a cylinder in terms of height (H) & diameter (D) instead of height & radius.

Common Conversions:

Tip: Generally we assume conversion factors are perfect numbers (infinite sig figs) unless otherwise noted.

$160\underline{9}$ m = 1 mi	12 in = 1 ft	60 s = 1 min	1000 g = 1 kg
2.54 cm = 1 in	1 cc = 1 cm^3 = 1 mL	60 min = 1 hr	100 cm = 1 m
1 cm = 10 mm	1 yard = 3 ft	3600 s = 1 hr	1 km = 1000 m
1 furlong = 220 yards	$528\underline{0}$ ft = 1 mi	24 hrs = 1 day	180° = π rad
		1 fortnight = 14 days	

Example 1:

Convert 23 in to miles and write your answer *with appropriate prefix* and correct sig figs. Note: we typically assume all of the above conversion factors have infinite sig figs (with the exception of $160\underline{9}$ m = 1 mi).

$$\frac{\underline{23}\text{ in}}{} \times \frac{2.54\text{ cm}}{1\text{ in}} \times \frac{1\text{ m}}{100\text{ cm}} \times \frac{1\text{ mi}}{160\underline{9}\text{ m}} = 0.0003\underline{6}3\text{ mi} = 3\underline{6}3\ \times 10^{-6}\text{mi} = 360\ \mu\text{mi}$$

Example 2:

As part of the problem statement in an exam question, your instructor tells you to assume 1 lbs = $2.\underline{2}$ kg. You are asked to use this (and your conversions listed above) to convert $234\frac{\text{kg}}{\text{m}^2}$ to PSI or $\frac{\text{lbs}}{\text{in}^2}$. Write the final *answer in scientific notation* with correct sig figs.

$$\frac{23\underline{4}\text{ kg}}{\text{m}^2} \times \frac{1\text{ lbs}}{2.\underline{2}\text{ kg}} \times \frac{1^2\text{ m}^2}{100^2\text{ cm}^2} \times \frac{2.54^2\text{ cm}^2}{1^2\text{ in}^2} = 0.06\underline{8}6\ \frac{\text{lbs}}{\text{in}^2} = 6.\underline{8}6\ \times 10^{-2}\text{ PSI} = 6.\underline{9}\ \times 10^{-2}\text{ PSI}$$

Density in physics given by $\rho = \frac{m}{V}$ and typically has units $\frac{\text{kg}}{\text{m}^3}$ or $\frac{\text{g}}{\text{cm}^3}$. Density can effectively be used to convert mass to volume and vice versa.

1.18 Two accelerations are given as $a_1 = 0.060\frac{\text{mm}}{\text{s}^2}$ and $a_2 = 49500\frac{\text{ft}}{\text{hr}^2}$

a) Determine the number of sig figs in each acceleration.

b) Convert a_1 to $\frac{\text{ft}}{\text{hr}^2}$ units. Write your final answer with correct sig figs and *scientific* notation.

c) Convert a_2 to SI units (m and s). Use correct sig figs & *engineering* notation with appropriate prefix.

1.19 Convert $6\underline{0}$ mph to m/s.

1.20 A space rock composed of aluminum and iron has a density of $1.010 \times 10^2\frac{\text{g}}{\text{in}^3}$. Convert to $\frac{\text{kg}}{\text{m}^3}$. Write your answer with correct sig figs and proper scientific notation. *Challenge:* What % of the rock (by volume) is aluminum?

1.21 Near earth's surface we assume 1 lbs = $453.\underline{6}$ g. Convert $13.5\frac{\text{lbs}}{\text{day}^2}$ to SI units (kg and s). Write your final result with correct sig figs in *scientific* notation.

1.22 Convert $12.3\frac{\text{in}}{\text{hr}^2}$ to SI units (m and s). Use correct sig figs & *engineering* notation with appropriate prefix. Note: the units of your final answer should look something like $\frac{\text{km}}{\text{s}^2}, \frac{\mu\text{m}}{\text{s}^2}$, etc.

Using density to relate mass and volume

In most physics classes we use the symbol ρ (pronounced "rho") to represent density. The density equation is thus

$$\rho = \frac{m}{V}$$

Obviously this can be rearranged as needed to determine mass or volume. Another way to think of it is density can be used as a conversion factor between mass and volume.

1.23 I have 4.0 cm^3 of aluminum with density 2.7 $\frac{\text{g}}{\text{cm}^3}$. What is the equivalent mass?

1.24 I have 4.2×10^2 kg of aluminum with density 2700 $\frac{\text{kg}}{\text{m}^3}$. What is the equivalent volume?

Using volume & mass flow rates

"Volume Flow Rate" means $R_{vol} = \frac{Volume}{time} = \frac{V}{t}$ and has units of $\frac{\text{m}^3}{\text{s}}$ (another common unit is $\frac{\text{Gallons}}{\text{minute}}$)

"Mass Flow Rate" means $R_{mass} = \frac{mass}{time} = \frac{m}{t}$ and has units of $\frac{\text{kg}}{\text{s}}$

Notice we can convert from mass flow rate to volume flow rate using density as above.

To convert a mass to a volume we divide by density (as in **Example 4** above). To convert R_{mass} to R_{vol} we similarly divide by ρ! Cool.

1.25 At one point during a flood, the peak water flow rate at the Oroville Dam spillway was 100,000 $\frac{\text{cubic feet}}{\text{second}}$. Determine the amount of $\frac{\text{kg}}{\text{day}}$ flowing over the dam. You may assume the density of water is 1000 $\frac{\text{kg}}{\text{m}^3}$. **Round your final answer to three sig figs and answer with correct scientific notation.**

1.26 Putting it all together: A nut is modeled as a rectangular solid with a cylindrical hole drilled out of it. This assumes we are ignoring the threads. A sketch of such a nut is shown at right. Note: figure not to scale.

a) Determine the volume of metal in the nut in terms D, s, and t.

b) You are told the nut is made of gold with density 19.3 $\frac{\text{g}}{\text{cm}^3}$. The nut is $4.50\ \times 10^{-3}$ m thick with side $s = 2.00$ in and hole diameter 50.3 mm. Determine the mass of the nut in kg. Write your answer with correct sig figs and scientific notation.

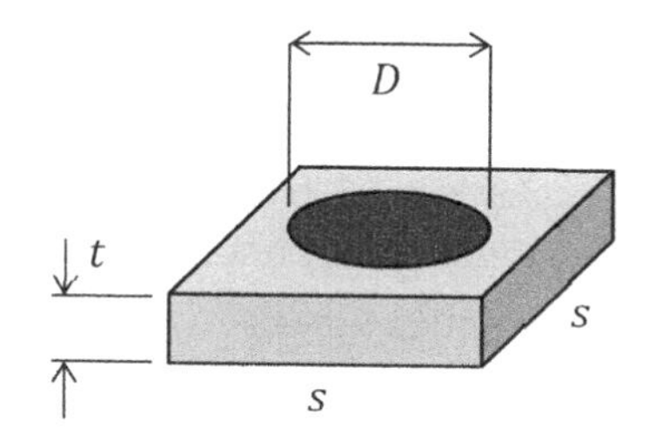

1.27 Assume a nucleus is a sphere of diameter 12 μÅ. You are told 1 Å = 10^{-10} m. Determine the surface area of the nucleus in fm^2. Write your answer with correct sig figs.

1.28 A plate has a square hole cut from it. The circle has diameter D and the square hole has side length $\frac{D}{3}$. The mass of the plate, *after the square piece is removed*, is m. The thickness of the plate is t. Figure not to scale. Determine an algebraic expression for the density. Write the answer as a number with three sig figs times $\frac{m}{D^2t}$.

1.29 Many feel this is tricky: Consider the hexagonal nut shown in the figure. The nut is a regular hexagon with a circular hole drilled in it. For this problem ignore the threads inside the hole. Suppose each side of the hexagon is s = 6.0 mm long. The hole has diameter of D = 10.7 mm. The nut is t = 2.0 mm thick. The density of aluminum is 2.7 g/cm^3.

a) Determine the volume of the nut *algebraically* (in terms of the variables before you plug in any numbers).

b) Determine mass of the nut. Get the sig figs correct & answer in engineering notation. Note: this is <u>not</u> a realistically sized nut…try sketching it to size to see what I mean.

1.30 A spherical drop of oil is dropped onto the surface of a smooth pond. Within a matter of seconds after touching the surface, the oil rapidly spreads out into the shape of a pancake (very short cylinder). The oil slick formed on the surface eventually ends up having a thickness roughly equal to a single molecule of oil! Let's use a demo to figure out thick the oil molecule is.

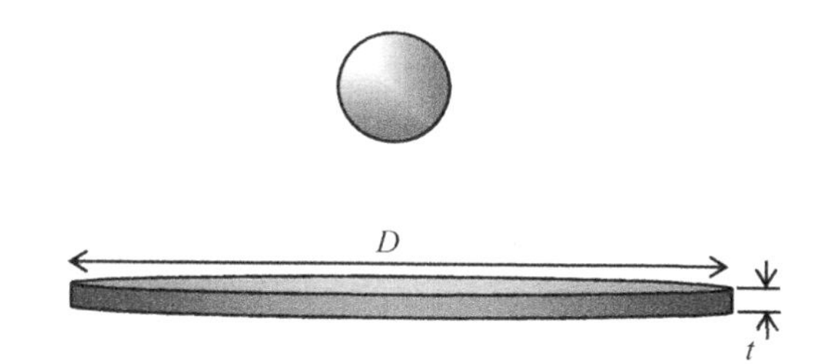

a) Determine the diameter of the initial drop.

b) Determine an expression for t in terms of the other variables.

c) Use values provided by your instructor during the demo to determine the thickness. As a back-up, I found a video where a 0.5 μL (yes, that's micro!) drop created an oil slick with diameter 60 cm.

One can learn more about this by doing a web search for "Benjamin Franklin oil drop". I learned of a scientist named Agnes Pockles (1862-1935). She was denied higher education. In her kitchen she created a device for performing this type of experiment with much greater precision. Her device is the precursor to the Langmuir trough which is still used in biophysics labs today. Her work was published in the prestigious journal *Nature* with the help of Nobel Laureate Lord Rayleigh. She had numerous other publications/contributions and was a pioneer of the field of surface science. At age 70 she was awarded an honorary doctorate.

1.31 A 45-45-90 triangular plate has a circular hole cut from it. The two sides of the right triangle are length s while the hole has diameter $\frac{s}{4}$. The mass of the plate, *after the circular piece is removed*, is M. The thickness of the plate is x. Figure not to scale. Determine an algebraic expression for the density. Write the answer as a number with three sig figs times $\frac{M}{s^2x}$.

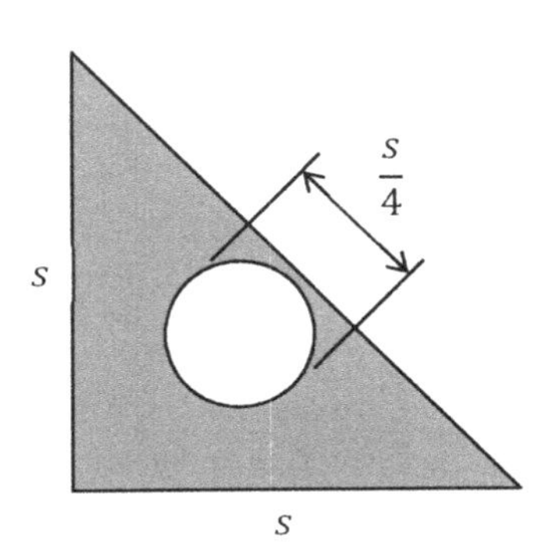

1.32 This problem was inspired by spherical ice ball makers which seem to be all the rage these days. We will consider a simplified model of such a device.

A cylinder of brass has diameter 4.00 inches, height 2.40 inches and density $8.73\frac{\text{g}}{\text{cm}^3}$. A hemisphere of radius R must be removed from the cylinder to allow space for the ice. *After the hemispherical hole is made*, the desired weight is 5.55 lbs. Near earth's surface humans typically assume 1 lbs = $0.453\underline{6}$ kg.

Determine the radius of the hemispherical hole which gives the desired weight of 5.55 lbs. **Answer in units of meters with correct sig figs and scientific notation.** I will also accept engineering notation with appropriate prefix in lieu of scientific notation.

I want to cut a hemispherical hole. *After the hole is cut*, the weight is 5.55 lbs. Figure out the radius of the hemispherical hole.

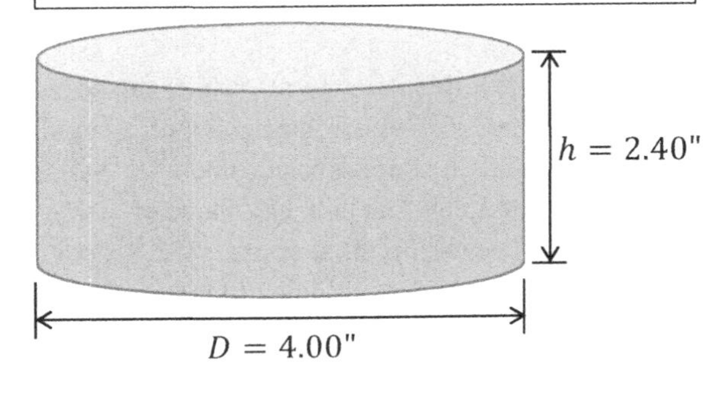

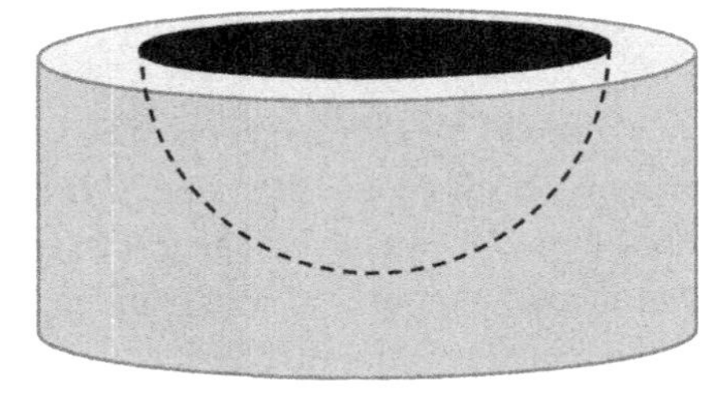

$[\mathrm{M}] = \text{mass} = \mathrm{kg}$ $[\mathrm{L}^2] = \text{area} = \mathrm{m}^2$ $[\mathrm{T}] = \text{time} = \mathrm{s}$ $\left[\frac{\mathrm{L}}{\mathrm{T}^2}\right] = \text{acceleration} = \frac{\mathrm{m}}{\mathrm{s}^2}$

$[\mathrm{L}] = \text{length} = \mathrm{m}$ $[\mathrm{L}^3] = \text{volume} = \mathrm{m}^3$ $\left[\frac{\mathrm{L}}{\mathrm{T}}\right] = \text{velocity} = \frac{\mathrm{m}}{\mathrm{s}}$ $\left[\frac{\mathrm{L}\cdot\mathrm{M}}{\mathrm{T}^2}\right] = \text{force} = \frac{\mathrm{kg}\cdot\mathrm{m}}{\mathrm{s}^2} = \mathrm{N}$

Be careful. When you write v it doesn't mean the same thing as $[v]$.

- Here v is the variable speed which includes both a number and units when you substitute in a value.
- Here $[v]$ implies you should write only *the units of speed.*
- After you write the units of speed, the square brackets disappear.
- Notice the following tricky points:
  - The variable m (italicized) implies mass which has *units* of kg
  - The unit m (not italicized) stands for meters, the SI unit of length
  - $[m] = \mathrm{kg}$ while $[x] = \mathrm{m}$

*****A handout with explanations & worked examples can be found online.*****
The handout is linked next to the Chapter 1 Solutions link online (in the comments sections).

1.33 Use this first problem as the example. Blatantly follow the solution if desired. In the following equations L is radius, v is velocity, V is volume, m is mass, and g is the magnitude of acceleration due to gravity. Consider the following equations and determine the units of the variable specified.

a) $K = \frac{1}{2}mv^2$. Find the units of K.

b) $\rho = \frac{m}{V}$. Find the units of ρ.

c) $P_D = \frac{1}{2}\rho v^2$. Find the units of P_D.

d) $\omega = \sqrt{\frac{g}{L}\tan\theta}$. Find the units of ω.

1.34 Consider the following set of measurements shown in the table. Notice the units are given for each measurement in parentheses in the column headings.

t (s)	m (kg)	M (kg)	x (m)	r (m)	s (m)	v (m/s)	a (m/s^2)	F (N)
1.00	2.50	8.1	4.200	1.00	6.28	5.39	6.4	16.5

a) A force equation involving drag is given by

$$ma = -bv^2$$

Determine the appropriate units for the constant b. Answer only in SI units (s, kg, & m in this case).

b) An equation for the force due to gravity is given by

$$F = \frac{GmM}{r^2}$$

Determine the appropriate units for "big G". Answer only in SI units (s, kg, & m in this case).

c) Explain why the following formula cannot be correct. Hint: first determine the units of each term on the right side and compare to the units of x.

$$x = s\frac{at}{v} - t\sqrt{\frac{m}{F}}$$

d) A common formula in oscillations is given by

$$x = A\cos\omega t$$

where ω is angular frequency, A is amplitude, t is time, and x is distance from equilibrium. Determine the units of A and ω. Hint: In general, functions and their arguments have no units. We say the arguments of trig functions are radians, but radians are simply a place holder unit.

e) A common formula in circuits is given by

$$Q = Q_{max}e^{-\alpha t}$$

What are the units of α? Think: do the units of Q and Q_{max} matter?

1.35 An equation for speed is determined as $v = \sqrt{\frac{rg}{\mu_s}}$ where g is the magnitude of the acceleration due to gravity and r is a radius. Determine the units of μ_s.

1.36 Suppose you have a force equation given by

$$F = ax - bx^3$$

where force (F) is measured in Newtons and distance (x) is in m. Recall, $1\text{ N} = 1\frac{\text{kg·m}}{\text{s}^2}$. Determine the units of the constants a and b.

1.37 Suppose you have a force equation given by

$$F = cv + dv^2$$

where force (F) is measured in Newtons and speed (v) is in $\frac{\text{m}}{\text{s}}$. Recall, $1\text{ N} = 1\frac{\text{kg·m}}{\text{s}^2}$. Determine the units of the constants c and d.

1.38 For now, assume pressure (P) is measured in Pascals (Pa). Notice that the variable P is italicized but the units Pa are not. Note: you will learn in chapter 14 $1\text{ Pa} = 1\frac{\text{N}}{\text{m}^2}$. Recall also $1\text{ N} = 1\frac{\text{kg·m}}{\text{s}^2}$. Assume volume ($V$) is measured in m^3. Temperature is measured in kelvin (K). Number of moles (n) has units of moles (mol).

a) Perhaps you have heard of the ideal gas law ($PV = nRT$). Determine the units on the constant R.
b) There is another gas law which provides a better model for certain situations called "the van der Waals equation of state". The equation of state is

$$\left[P + a\left(\frac{n}{V}\right)^2\right]\left(\left(\frac{V}{n}\right) - b\right) = RT$$

Determine the units of the new constants a and b.

Note: I read somewhere once that before 1968 *degrees* kelvin was the norm but I guess you shouldn't say the degrees anymore. Many capitalize Kelvin as it comes from a name. I don't really care if you capitalize or not but in formal writing perhaps you would look up appropriate usage in the NIST style guide?

Note[2]: if you use chemistry units for pressure and volume (atm and L respectively) books sometimes use r (not R).

Note[3]: If you set the constants equal to zero in the van der Waals state equation to zero you get back the ideal gas law.

1.39 Physics & Aerospace Majors: Force is measured in Newtons (N) where 1 N = 1kg·m/s². From experience we expect the force keeping an object in circular motion should increase with increasing speed and mass but also with decreasing radius. If we model the force as

$$F = \frac{m^k v^n}{r^q}$$

where k, n, and q are positive exponents we know our model will fit our expectations. If only we could figure out the exponents. Use unit analysis to determine the correct exponents of mass, speed, and radius.

1.40 Physics & Aerospace Majors: Consider the simple pendulum shown at right. A simple pendulum consists of a small mass tied to the end of a string that oscillates back and forth along the curved path shown. If the diameter of the mass is much smaller than the length of the string, we may treat the mass as a point mass (mass with no size at all). We assume the string has negligible mass compared to the point mass. We also assume the string doesn't stretch (inextensible string).

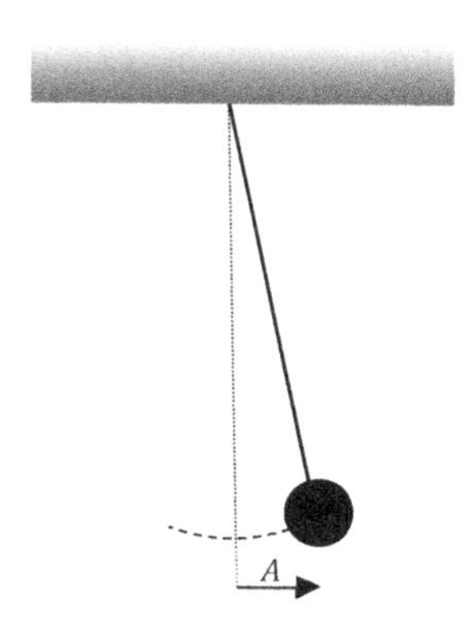

The time for the mass to swing back and forth one time is called the period (T) of oscillation. Without knowing anything more, we might suspect the T to depend on the length of the string L, the mass of the sphere m, the amplitude of the oscillation A, and the magnitude of the acceleration due to gravity (g). Said another way, we expect

$$T = kL^a m^b g^c A^d$$

where a, b, c, & d are exponents (positive or negative) and k is a *dimensionless* (aka unitless) constant. Note: amplitude is the distance the ball travels. Doing a brief experiment in front of the class reveals that, for small angles, the amplitude doesn't affect the period! Removing this term, by setting $d = 0$, gives

$$T = kL^a m^b g^c$$

Use dimensional analysis to determine the other exponents. You might be surprised at the results.

1.41 Physics & Aerospace Majors: In fluid dynamics it is of interest to study a dimensionless parameter known as Reynolds number (Re). For a sphere moving in a fluid, the following parameters might be of interest:

1) ρ = density of the fluid
2) D = diameter of the sphere
3) μ = Greek letter "mu" = dynamic viscosity of the fluid $\left(\text{units of } \frac{\text{kg}}{\text{m}\cdot\text{s}}\right)$
4) v = speed of the sphere

To incorporate the above elements, let us assume

$$Re = \rho^1 D^a \mu^b v^c$$

where a, b, and c are exponents (could be positive or negative). The exponent of ρ is set to 1 by convention. Use dimensional analysis to determine the exponents of each term. It turns out Re is extremely useful in fluid mechanics. Do a web search for "video Reynolds number" to see some amazing stuff.

1.42 Late in Chapter 6 will derive a formula $\psi = \sqrt{g \tan\theta(R + L\sin\theta)}$ where R is a radius, L is the length of a string, θ is the angle made by the string, and g is the magnitude of the acceleration due to gravity. Determine the units of ψ. This symbol is the Greek letter psi, pronounced like "sigh".

1.43 You are told an equation exists such that

$$\frac{x}{v^3} = k\sqrt{\frac{a}{rt^2}}$$

where x is position (same units as distance), v is velocity, t is time, a is acceleration, r is a distance, and k is an unknown constant. Determine the units of the unknown constant k. The answer may look pretty unusual as I made up the craziest equation I could think of to give you some hard practice.

1.44 Most of the following terms are actually used in physics problems at some point during your first semester. Which of the following terms have matching units? There may, or may not be, more than one match. Assume that m is mass in kg, h & r are distances in m, v is speed in m/s, a & g are accelerations in m/s^2, and t is time in seconds.

ma	mgh	mvt	mvr	$\frac{1}{2}mv^2$	$m\frac{v^2}{r}$	mht

1.45 In the solutions I made a set of instructions with screen captures which should help you make plots. Open the solutions and follow the instructions before attempting the next two problems.

1.46 Do after completing **1.45**. Later in the semester we encounter some tricky equations. Determine θ.

$$2\cos\theta - 2.232 + \sin\theta = 0$$

1.47 Do after completing **1.45**. In a problem relating to diffraction one must solve the transcendental equation

$$\frac{\phi}{\sqrt{2}} = \sin\phi$$

where ϕ is in radians. This equation determines the angle for full width at half maximum (FWHM) of the central bright fringe of a single slit diffraction pattern. Solve the equation for ϕ.

Absolute, Fractional and Percent Errors

Suppose you measure a distance as $x = 2.0$ cm and you expect to be off by no more than half of the rightmost digit.

Absolute error in x is $\delta x = \frac{1}{2}(rightmost\ digit) = \frac{1}{2}(0.1) = 0.05$ cm.
The symbol δ is the lowercase delta from the Greek alphabet.
While conventions in other books vary, many assume δ indicates absolute error.

Fractional error is given by $\frac{\delta x}{x} = \frac{0.05\text{ cm}}{2.00\text{ cm}} = 0.025 \approx 0.03$.
Notice fractional error has no units.
Common convention is to round errors to 1 sig fig.
Exception: if the first digit is 1, round to 2 sig figs.

Percent error is simply expressing the fractional error as a percent.
Percent error is thus $\frac{\delta x}{x} \times 100\% = \frac{0.05\text{ cm}}{2.00\text{ cm}} \times 100\% = 0.025 \times 100\% \approx 3\%$
Percent error is a good way to compare the quality of measurements with different units.

1.48 Determine the *percent* error in each of the following measurements. To be clear, the $\pm$ number in each measurement is the *absolute* error.

a) $\rho = 50.1 \pm 0.1 \frac{\text{g}}{\text{cm}^3}$
b) $x = 3.0 \pm 0.1$ m
c) $d = 1.0 \pm 0.3$ m
d) $T = 0.060 \pm 0.001$ s
e) $t = 6\underline{0} \pm 1$ s
f) $f = 9.0 \pm 0.5 \frac{\text{m}}{\text{s}^2}$
g) $g = 9.5 \pm 0.5 \frac{\text{m}}{\text{s}^2}$
h) $h = 10.5 \pm 0.5 \frac{\text{m}}{\text{s}^2}$
i) $i = 19.5 \pm 0.5 \frac{\text{m}}{\text{s}^2}$
j) $j = 1\underline{0} \pm 1 \frac{\text{m}}{\text{s}^2}$

1.49 The diameter of regulation table tennis balls must be $D = 40.0 \pm 0.5$ mm. The mass of regulation balls must be $m = 2.70 \pm 0.03$ g.

a) Determine the volume of a 40.0 mm ball. Express with correct sig figs, scientific notation, and units of m^3.

b) Determine the surface area of 40.0 mm ball. Express with correct sig figs, scientific notation, and units of m^2.

c) The ball is a celluloid spherical shell with thickness t. The celluloid has density ρ. Determine an expression for the thickness of the shell in terms of ρ, m and D.

d) Assume the celluloid used to make the balls has density $1375 \frac{\text{kg}}{\text{m}^3}$. Determine the range of thicknesses of a regulation ball. Express your answer as $t = \# \pm \#$ and as $t = \# \pm \%$ in units of mm similar to the diameter specification above. Assume we may ignore the mass of the air inside the ball.

e) Assume the air inside the ball has density $1.29 \frac{\text{kg}}{\text{m}^3}$. Determine the mass of the air inside the ball. Express your answer as a percentage of the ball's mass ($m = 2.70$ g). Notice this mass is roughly equal to the uncertainty in the ball's mass. It was probably reasonable to ignore it in our previous calculation.

f) **Challenge**: In part c it is easiest to use the volume of plastic used in the ball is

$$V = (Surface\ Area) \times thickness = (S.A.) \times t$$

A more precise method uses $V = V_{outer} - V_{inner}$ where V_{outer} uses diameter D and V_{inner} uses diameter $D - 2t$. Determine the difference in calculated volume between the methods in terms of D and t.

g) **Challenge:** for what thickness does the approximation $V = (S.A.) \times t$ cause a 1% error in the volume calculation. Answer as a fraction times D.

Note: perhaps you are wondering if the air inside the ball affects the scale reading of the ball when it is placed on a mass balance. We will learn about this when we discuss buoyant forces much later...

1.50 Disclaimer: A did a quick search for specs on PVC. The two main types appear to be Schedule 40 and Schedule 80. Depending on which source you use you might find slightly different numbers. That said, any numbers in the ballpark should serve our purpose just fine here. A Schedule 40 PVC pipe has density ρ, outer diameter D, inner diameter d and height h.

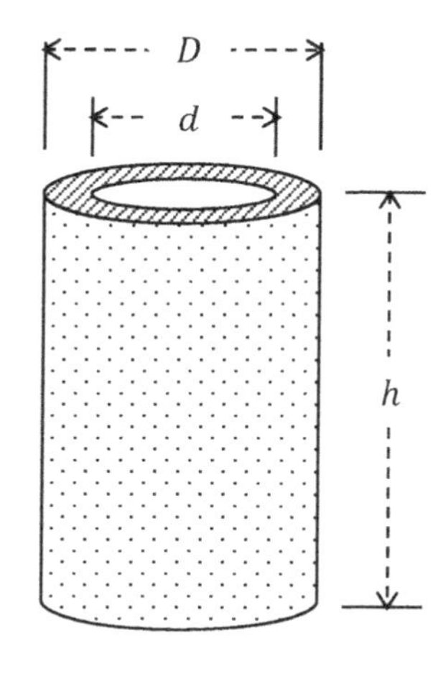

a) Determine an expression for the surface area of the sidewall of the pipe. The sidewall is indicated by sparse dots in the figure.

b) Determine an expression for the surface area of the end. The end is shaded with diagonal lines in the figure.

c) Determine an expression for the mass of the pipe.

d) For 1-inch nominal pipe, one website stated $D = 1.315$ in, $d = 1.029$ in, and $\rho = 1295 \frac{\text{kg}}{\text{m}^3}$. Nominal means "size in name only"; a 1-inch nominal pipe is approximately but not exactly 1 inch in diameter. Determine the mass *per unit length* of such a pipe. Answer with correct sig figs in scientific notation using $\frac{\text{g}}{\text{cm}}$.

e) In most calculations we would probably ignore the area of the two ends of the pipe compared to the sidewall. If the pipe is short, this will introduce errors in our calculations. What length of 1-inch pipe has the combined area of both end caps equal to a 1% correction to the sidewall area?

1.51 Read about order of magnitude calculations (Fermi calculations). Do the following estimations without the internet.

a) Estimate the number of heartbeats in your lifetime.
b) Estimate the maximum speed of a garden variety snail in furlongs per fortnight. Note: one furlong is an eighth of a mile. A fortnight is a unit of time equal to two weeks.
c) Estimate how many gallons of gasoline are required to drive a car across the country. I'm assuming it is not electric…
d) Estimate how much water you use per day.
e) Estimate the volume of air in a room then use that to estimate the number of air molecules in that room.
f) Estimate the time in your life spent at stop lights.
g) Estimate the time in your life you will waste on paperwork…or perhaps some things are better left unknown.
h) Estimate the number of stars you can see on a clear night.
i) Estimate the area of your college campus…or the number of bathrooms on campus.

Obscure comment: On occasion the sig fig rules give answers that vary depending on the order of operation! This serves as a reminder to us that sig fig rules are merely an estimate of precision. By using the sig fig rules we will be in the right ballpark but might not always have the best estimate of precision. In lab you will learn more about error analysis. This is a skill that takes many years to master…

Example: Suppose you are given two spheres with radii 9.6 cm and 5.2 cm respectively. You, Rogelio, and Vega are asked to find the combined volume of the two spheres. You are each told to keep track of the sig figs.

You decide to the problem as follows:

$$\begin{aligned} V_{tot} &= V_1 + V_2 = \frac{4}{3}\pi r_1^3 + \frac{4}{3}\pi r_2^3 \\ &= 4.18\underline{9} \times 9.\underline{6}^3 + 4.18\underline{9} \times 5.\underline{2}^3 \\ &= 3\underline{7}06 + 5\underline{8}9 \\ &= 4\underline{2}95 = 4.3 \times 10^3 \text{cm}^3 \end{aligned}$$

Rogelio decides to do the problem by first factoring out the constants like this:

$$\begin{aligned} V_{tot} &= V_1 + V_2 = \frac{4}{3}\pi r_1^3 + \frac{4}{3}\pi r_2^3 = \frac{4}{3}\pi(r_1^3 + r_2^3) \\ &= 4.18\underline{9}(9.\underline{6}^3 + 5.\underline{2}^3) \\ &= 4.18\underline{9}(8\underline{8}5.7 + 1\underline{4}0.6) \\ &= 4.18\underline{9}(10\underline{2}5.3) \\ &= 42\underline{9}5 = 4.30 \times 10^3 \text{cm}^3 \end{aligned}$$

Finally, Vega determines an estimate of precision based on percent errors. She first makes the assumption that when someone measures 9.6 cm they are implying 9.6 ± 0.05 cm. Similarly, she assumes that the other measurement was 5.2 ± 0.05 cm. This implies the percent error on each measurement is given by:

$$\%\text{err in } r_1 = \frac{0.05}{9.6} \times 100\% = 0.52\%$$

$$\%\text{err in } r_2 = \frac{0.05}{5.2} \times 100\% = 0.96\%$$

Since each radius was cubed, Vega expects the error for each radius to contribute three times. Then Vega further takes the cautious approach and assumes the worst case scenario; Vega assumes that the error in measuring one radius will not cancel out the error in measuring the other radius. For example, Vega is assuming that either both radii were measured too large or too small (not one of each). This gives a total error as follows:

$$\begin{aligned} \%\text{err} &= 3(\%\text{err in } r_1) + 3(\%\text{err in } r_2) \\ &= 3(0.52\%) + 3(0.96\%) \\ &= 1.56\% + 2.88\% \\ &= 4.44\% \approx 4\% \end{aligned}$$

Using this method, Vega decides that the total volume is given by

$$4295 \pm 4.44\% = 4295 \pm 190 \approx 4295 \pm 200$$

Vega infers the 3rd digit isn't significant and the result should be written as $4.3 \times 10^3 \text{cm}^3$.

1D MOTION

Position Vector = $\vec{x}$ or $\vec{r}$ =location (magnitude & direction) relative to the origin

Displacement Vector = $\Delta\vec{x}$ or $\Delta\vec{r}$ =CHANGE in position (magnitude & direction)

Distance Scalar = Δx or Δr **WATCH OUT!** If the object changes direction $\Delta x \neq \|\Delta\vec{x}\|$!

Instantaneous Velocity Vector = $\vec{v} = \frac{d\vec{x}}{dt}$ or $\frac{d\vec{r}}{dt}$ includes magnitude & direction

Average Velocity Vector = $\vec{v}_{avg} = \frac{\Delta\vec{x}}{\Delta t}$ or $\frac{\Delta\vec{r}}{\Delta t}$ includes magnitude & direction

Instantaneous Speed = $v = \|\vec{v}\|$ =the magnitude only of $\vec{v}$

Average Speed = $v_{avg} = \frac{Total\ distance}{Total\ time}$ **WATCH OUT!** If the object changes direction $v_{avg} \neq \|\vec{v}_{avg}\|$!

Instantaneous Acceleration Vector = $\vec{a} = \frac{d\vec{v}}{dt}$ includes magnitude & direction

Average Acceleration Vector = $\vec{a}_{avg} = \frac{\Delta\vec{v}}{\Delta t}$ includes magnitude & direction

WATCH OUT! Even though the above definitions all have arrows, almost no physics book includes the arrows for 1D motion *equations*. For example, under constant acceleration displacement is given by

$$\Delta x\hat{\imath} = (v_{ix}t)\hat{\imath} + \left(\frac{1}{2}a_x t^2\right)\hat{\imath}$$

$$\Delta x = v_{ix}t + \frac{1}{2}a_x t^2$$

After cancelling the $\hat{\imath}'s$ you are supposed to know, in this instance, Δx implies *displacement*, not *distance*!!!

Usually we just assume right is the positive direction unless lots of things in the problem are going left.
Moving forward implies positive velocity. Moving backward implies negative velocity.
If velocity and acceleration have the same sign the object is speeding up (opposite signs then slowing down).

$v > 0$ AND $a > 0$	moving forward and speeding up
$v > 0$ AND $a < 0$	moving forward and slowing down
$v < 0$ AND $a < 0$	moving backward and speeding up
$v < 0$ AND $a > 0$	moving backward and slowing down

Be careful when an object reverses direction! At turnaround points the velocity is instantaneously zero but not necessarily the acceleration. Furthermore, when an object goes left then right (or up then down) the displacement will partially cancel while the two distances will not. Average velocity and average speed will not be equal in magnitude anymore!!! See below for an example.

The following equations are only valid for CONSTANT acceleration. For problems in which the acceleration changes you need to split the problem into separate parts such that each part has constant acceleration.

$$\Delta x = v_{ix}t + \frac{1}{2}a_x t^2 \qquad v_{fx}^2 = v_{ix}^2 + 2a_x\Delta x \qquad v_{fx} = v_{ix} + a_x t \qquad \Delta x = \frac{1}{2}(v_{fx} + v_{ix})t$$

People often write the kinematics equations the following way instead:

$$x_f = x_i + v_{0x}t + \frac{1}{2}a_x t^2 \qquad v_x^2 = v_{0x}^2 + 2a_x\Delta x \qquad v_x = v_{0x} + a_x t \qquad \Delta x = \frac{1}{2}(v_x + v_{0x})t$$

Notice that these equations are exactly the same if we make the identifications below:

$$\Delta x = x_f - x_i \qquad v_{fx} = v_x \qquad v_{ix} = v_{0x}$$

2.1 An object starts at the origin. It moves to the right 10.00 m in 4.00 s, then to the left 15.00 m in the next 1.00 s, then to the right 5.00 m in the last 1.00 s.
In this problem the symbol $\hat{\imath}$ means "to the right". Using $-\hat{\imath}$ means "to the left".
Note: ignore any vertical displacement; it was exaggerated in the figure to simplify labeling.
Note: ignore acceleration for this problem. Assume each stage of the journey occurs with constant speed.
Note: ignore any time required for reversing direction in this problem.

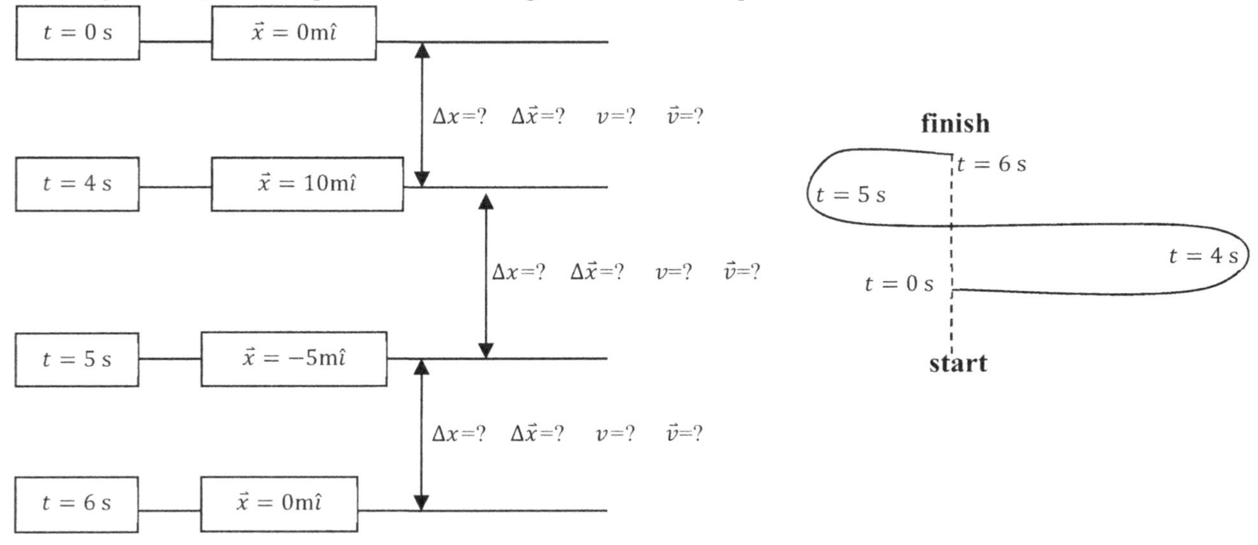

time interval	Δx (m)	$\Delta\vec{x}$ (m)	v (m/s)	$\vec{v}$ (m/s)
0-4 sec				
4-5 sec				
5-6 sec				
total trip (0-6 sec)				

Going forward, here is a problem solving strategy for 1D motion problems:

1. Read
2. Draw
3. Label diagram
4. Identify/list knowns/unknowns
5. Equations
6. Re-read to ensure you are heading in the right direction
7. Plug in any zeros then solve algebraically
8. Substitute in values if given
9. Check your result (units, reasonable, +/- is sensible, order of magnitude)

2.1¼ Suppose the landing speed for a huge jet airplane is about 170 miles per hour. Assume the plane is to slow to a stop at a rate of $0.175g$. Assume this rate is constant from the instant the plane touches down until the instant it comes to a full stop. Here g is the magnitude of freefall acceleration near earth's surface. Note: the numerical value is always $g = +9.8\frac{\text{m}}{\text{s}^2}$ (plus sign added for emphasis…g is always positive).

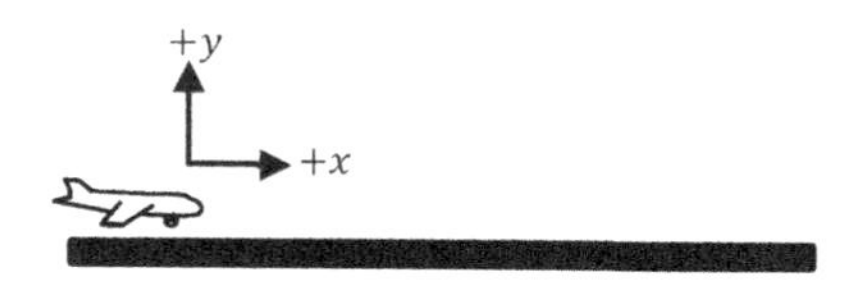

a) Is g gravity?
b) Is g the *acceleration* due to gravity?
c) Under what circumstances does $g = -9.8\frac{\text{m}}{\text{s}^2}$?
d) How far will the plane travel as it slows from touchdown to full stop? Answer in both meters and miles.
e) How long does it take to stop? Assume this is elapsed time between touchdown and stop.

2.1½ Perhaps you have seen those yellow barrels near a highway underpass designed to prevent death in the event of a car crash? It is my understanding they are called sand barrel arrays. To get a feeling for how these might work, let's assume the barrels cause a constant acceleration even though the acceleration actually changes quite a bit during the collisions. Assume a car comes in at $35.0\frac{\text{m}}{\text{s}} \approx 80$ mph and stops in a distance of 9.00 m.

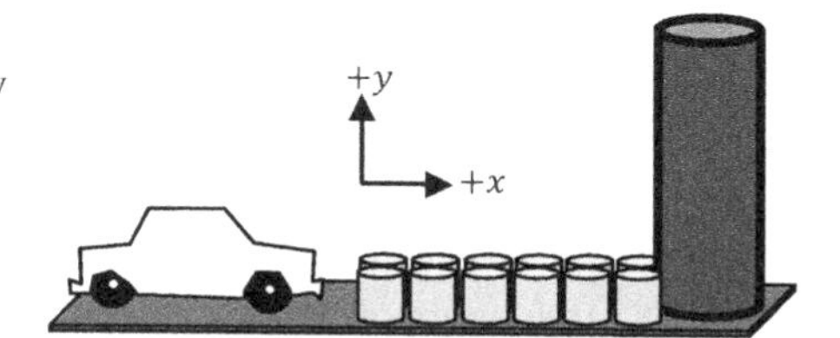

a) Determine the time to stop.
b) Determine the acceleration in $\frac{\text{m}}{\text{s}^2}$ using the coordinate system shown.
c) For comparison, assume no yellow barrels were present. If you hit the concrete you would come to a full stop in perhaps 1.5 m…approximately the distance between your face and the concrete structure at the moment your bumper impacts the concrete. Compute the acceleration for this case.
d) People like to use $g's$ when discussing acceleration. Convert your accelerations from the previous parts to $g's$ by dividing each term by g. Example: $a = 8.25\frac{\text{m}}{\text{s}^2}$ implies $\frac{a}{g} = 0.842$ implies $a = 0.842g$. Notice you don't have to include units when describing accelerations in terms of $g's$ (because the units are hidden inside… $g = 9.8\frac{\text{m}}{\text{s}^2}$).

Note: the acceleration a human can survive without death varies depending on several factors including the duration of acceleration and each particular human's body. Try a web search to learn more if you are interested.
Side note: In practice the barrels are not all filled with the same amount of sand. The first barrels to be impacted might be about half full while the barrels adjacent to the concrete structure are completely full. In real life we would not expect acceleration to be constant.

2.1¾ *Instructor tip:* use blue marker for first stage symbols and red marker for second stage symbols. A track star runs the 100-meter dash. She accelerated from rest to her top speed at a rate of $4.75\frac{\text{m}}{\text{s}^2}$ for the first 15.0 m of the race (figure not to scale). The runner finished the race at her top speed.

a) What is the top speed of the runner?
b) What was her time for the race?

Going Further: Create a spreadsheet to do the calculations for you. Then mess around with the value used for a until you get the current women's world record time of about 10.5 seconds.
Going Further: In real life, as you get going faster and faster your acceleration should decrease. A slightly better model might be accelerating at $5.5\frac{\text{m}}{\text{s}^2}$ for the first 7.5 m, then accelerating at $4.0\frac{\text{m}}{\text{s}^2}$ for the next 7.5 m. Try making a spreadsheet to compute the time of the race. See if the time is faster or slower than the initial conditions.

The goal of this page is to acclimate you to dealing with purely algebraic problems.
Why care? In general, we want to solve for our final results algebraically for a number of reasons.

1) Once you get good at it, it is faster.
2) It is easy to double check the units at the end. If the units *don't* match, we know we must go look for a flipped fraction, missed square root, or something similar in the algebra.
3) The final result is given in terms of initial parameters we can change. This allows us to test a wide variety of situations without having to redo the problem for every new set of initial conditions.
4) This type of solution can be used to create or check computer simulations.

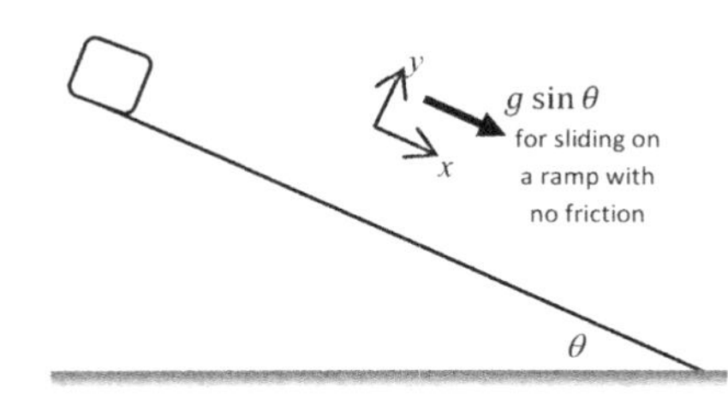

2.2 The figure shows a frictionless ramp. Assume the length of the ramp is L. Assume the box on the ramp is released from rest and experiences no friction. Assume the size of the box is negligible compared with the size of the ramp. The magnitude of the acceleration due to gravity in freefall is $g = 9.8\frac{\text{m}}{\text{s}^2}$. In all physics problems g is always positive. For a frictionless angled ramp the magnitude of acceleration is $g\sin\theta$.

a) Determine the angle required to reach the bottom of the ramp in time t. Answer in terms of L, g, and t.
b) If the ramp is 2.00 m long, what angle is required to make the object slide down the ramp in 1.40 sec?
c) Assuming the length is fixed, what angle will cut this time in half? Take a guess: do you think it would be more than, less than, or equal to 2θ? Now figure it out! Was your intuition correct?
d) Assuming the L is fixed, what angle gives the minimum time to go down the hill? Explain or derive the answer.

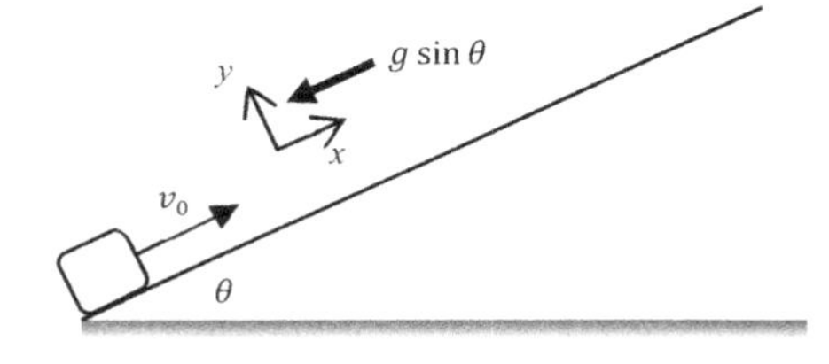

2.3 Now imagine a block is kicked and it slides up the ramp. Just after the kick, the block is moving up the ramp of length L with some unknown initial speed v_0. The acceleration (after the kick) is still $g\sin\theta$ directed down the ramp as shown in the figure. The block slides ¾ of the way to the top of the ramp before coming to rest when the angle is θ. Determine the initial velocity in terms of L, g, and θ. Use the coordinate system shown and watch your ± signs.

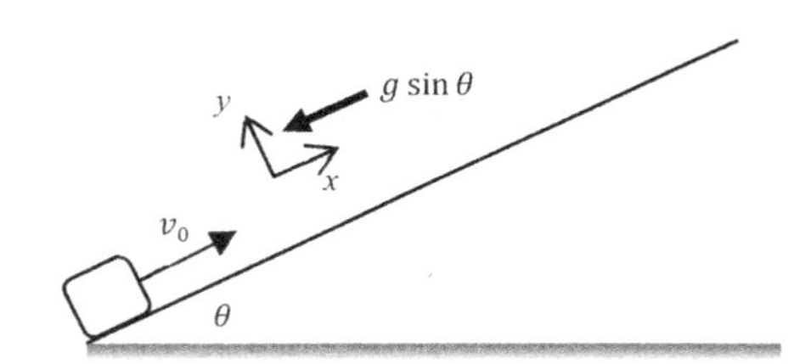

2.4 A block is kicked and slides up a ramp. Just after the kick, the block moves up the ramp with initial speed is 3.834 m/s. The acceleration (after the kick) is $g\sin\theta$ directed down the ramp as shown in the figure. The ramp length is 2.00 meters and the ramp angle is $\theta = 30.0°$.

a) What is the speed of the block after 1.00 sec?
b) How much *distance* is traveled by the block in 1.00 sec?

In a later chapter on rotational motion we will learn about rolling motion. For now, it suffices to know that the translational accelerations of various objects that roll *without slipping* down an incline of angle θ are given by

Solid Sphere	Solid Disk	Spherical Shell	Hoop/Thin Ring
$a = \frac{5}{7} g \sin\theta$	$a = \frac{2}{3} g \sin\theta$	$a = \frac{3}{5} g \sin\theta$	$a = \frac{1}{2} g \sin\theta$

Note: You are not expected to memorize these accelerations for testing purposes.

2.5 Suppose you have two identical length L ramps with the same angle θ. A solid sphere is placed on the first ramp while a spherical shell is placed on the second ramp. The solid sphere is released time Δt after the spherical shell. The two objects reach the end of the ramps at the same time.

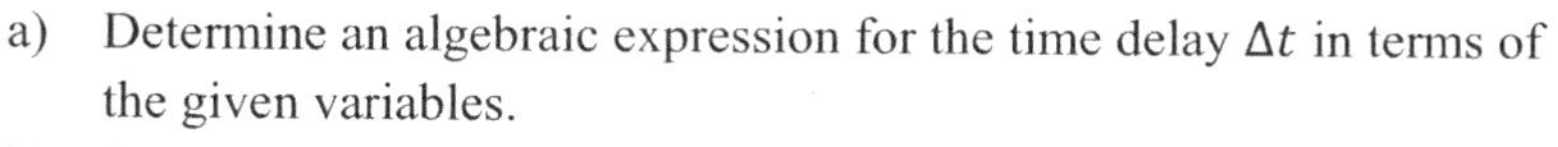

a) Determine an algebraic expression for the time delay Δt in terms of the given variables.

b) Assume $L = 2.00$ m and $\theta = 4.0°$. Determine the time for each ball to reach the end of the track and the time delay in seconds. Check with the data table below. In case you're wondering, I changed the angles to radians because Excel assumes the argument of all trig functions is radians. I used the RADIANS function. Try making the plot yourself for fun!

c) I made the data table shown below in hopes of finding a nice set of parameters to make a classroom demo. What are the pros and cons of large and small angles? Consider realistic factors like measuring the distance, times, and angles.

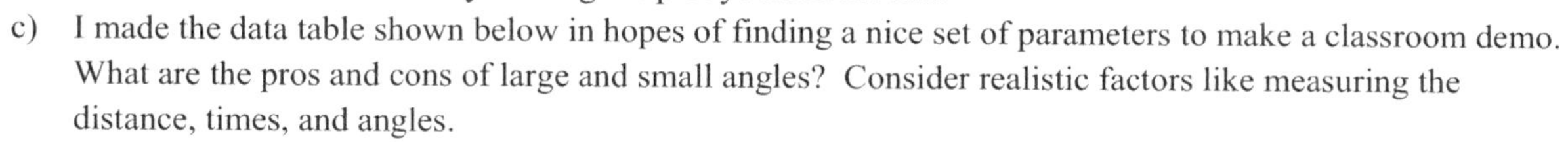

d) Notice I scratched out the row of calculations for $\theta = 90°$. Why is this model not valid at $\theta = 90°$?

e) **Challenge:** For most objects this model will not even work beyond 25° or 30°...*maybe* 45° for a rubber ball. Do you know what assumption in the model fails for these more moderate angles? We will discuss more in Chapter 10...

θ (deg)	θ (rad)	a_1 (m/s^2)	a_2 (m/s^2)	t_1 (s)	t_2 (s)	Δt (s)
0	0.000	0.00	0.00	#DIV/0!	#DIV/0!	#DIV/0!
1	0.017	0.12	0.10	5.72	6.24	0.52
2	0.035	0.24	0.21	4.05	4.42	0.37
3	0.052	0.37	0.31	3.30	3.61	0.30
4	**0.070**	**0.49**	**0.41**	**2.86**	**3.12**	**0.26**
5	0.087	0.61	0.51	2.56	2.79	0.23
6	0.105	0.73	0.61	2.34	2.55	0.21
7	0.122	0.85	0.72	2.17	2.36	0.20
8	0.140	0.97	0.82	2.03	2.21	0.18
9	0.157	1.10	0.92	1.91	2.09	0.17
10	0.175	1.22	1.02	1.81	1.98	0.17
20	0.349	2.39	2.01	1.29	1.41	0.12
40	0.698	4.50	3.78	0.94	1.03	0.09
~~90~~	~~1.57080~~	~~7.00~~	~~5.88~~	~~0.76~~	~~0.82~~	~~0.07~~

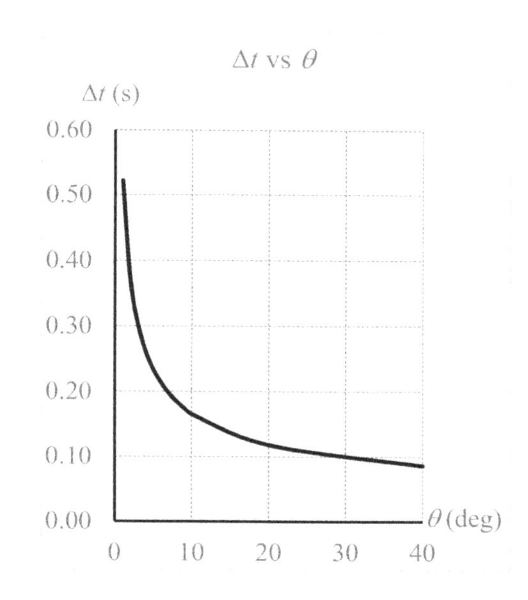

2.6 A driver is driving with constant speed v on a straight road. The driver sees a deer standing in the road a distance d away. After taking time t_r to react, he applies the brakes which cause the car to slow down at a constant rate. What minimum rate of slowing will cause the car to stop before hitting the deer? While the car approaches assume the deer remains in place like…like…like a deer in headlights.

2.7 A bullet enters a board travelling with speed v as shown in the figure. It exits the board travelling with 75% *less* speed. The thickness of the board is d. Determine the average acceleration of the bullet while it is in contact with the board. Note: any time you have bullet through board problems physicists typically make a few assumptions. Here are my assumptions; not necessarily the best but it gets a ballpark figure.

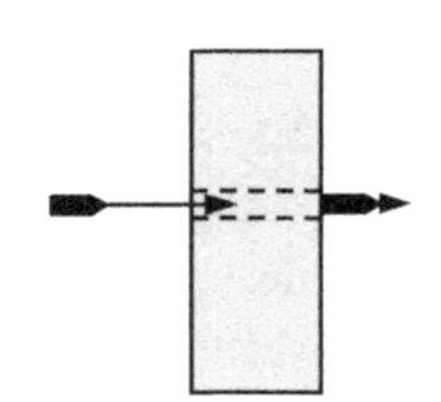

1) The board experiences negligible mass change
2) The bullet is indestructible (doesn't change shape)
3) The bullet is in considered "in contact" with the board over distance d.
4) Time to travel through board is time for tip of bullet to travel through the board.

2.8 When we say an object is in *freefall*, we assume gravity is the only force acting on object. Which of the following cases can be approximated well by freefall?

a) plane flies at constant altitude and constant speed
b) drop rock from a very *short* bldg
c) drop rock from a very *tall* bldg
d) the beginning of a rocket launch (thrusters on)
e) the moon orbits the earth

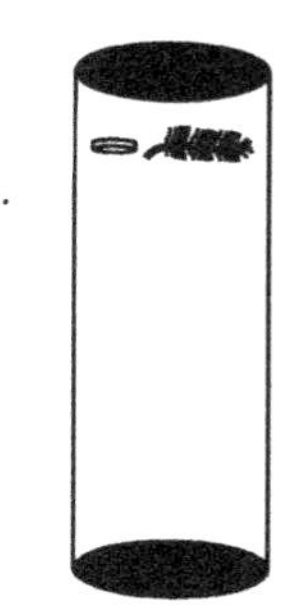

2.9 The coin and the feather demo. A coin and a feather are placed inside a tube that is sealed at both ends. The tube is evacuated to a very low air pressure using a vacuum pump. When the tube is suddenly turned upside down the coin and feather fall as if they are released simultaneously. Write down you observations. In particular, if air resistance can be considered negligible, do all objects in freefalling have approximately the same acceleration?

2.10 Basketball vs medicine ball demo. Your instructor climbs up on the desk and holds a basketball and a medicine ball about 3 meters above the floor. The balls are simultaneously dropped from rest.

a) Ignoring air resistance, which ball should reach the ground first? Explain.
b) A freefall kinematics model predicts the balls reach a top speed of about 17 miles per hour. Do you believe air resistance plays a huge factor in this problem? We learn more about air resistance after forces.
c) **Challenge:** In real life, which ball (if any) experiences a larger drag force just before impact? Assume the balls have identical radius.

2.11 & 12 Try using a simulation. I recommend looking at my supplemental handout on robjorstad.com…might need updating as simulation has changed.

2.13 When you throw a ball upwards it is in freefall (if air resistance is negligible). Think: after the ball leaves your hand you no longer exert any force on it; gravity is the only force (again, as long as air resistance is negligible). Assume you launch a ball by throwing it straight up. It leaves your hand approximately 2.00 m above the ground. It flies up to a max height of 6.00 m above the ground. Then it returns to earth.

a) Sketch a picture showing the approximate path taken by the ball.
b) What initial velocity is required for the ball to reach the max height stated in the problem?
c) What time is required to reach this max height?
d) What time is the entire time of flight?
e) Is total time of flight equal to twice the time to the max height?
f) What is the ball's impact velocity? Impact velocity is the velocity just before it reaches the ground.
g) What *distance* is traveled by the ball in 1.25 sec?
h) What is the ball's *displacement* after 1.25 sec?
i) Assume the origin of our coordinates system for this problem is at your feet. Said another way, assume ground level is the vertical postion $y = 0$. What is the *position* of the ball at 1.25 sec?
j) What is the *total distance* traveled by the ball?
k) What is the *total displacement* of the ball?
l) What is the *final position* of the ball?
m) Convert the impact speed to miles per hour. Think about this speed and discuss if you think air resistance is negligible. Or try a simulation online…

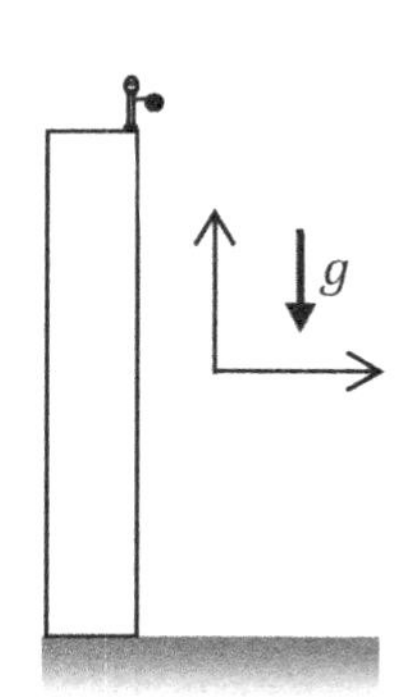

2.14 Thor drops *the* hammer from a tall building of height h. Thor's height is negligible compared to the entire building. Assume air resistance is negligible for the entire flight. Assume the magnitude of the acceleration due to gravity is g.

a) According to the coordinate system shown, which of the following equations is incorrect? Explain why?

$$\Delta y = v_{iy}t - \frac{1}{2}a_y t^2 \qquad \Delta y = v_{iy}t - \frac{1}{2}gt^2 \qquad y_f = y_i + v_{iy}t + \frac{1}{2}a_y t^2$$

b) Determine the time required to impact the ground. This is often called the time of flight.
c) Physicists often ask for the final velocity. This usually implies the impact velocity. The impact velocity of the hammer is the velocity just before it reaches the ground. Determine the impact velocity.
d) Is the impact *velocity* the same as the impact *speed*? Explain.
e) Determine the distance traveled by the hammer and its speed *after the first half of the total travel time.*
f) Determine the *time to travel half the distance* and speed of the hammer after the first half of the distance.

The purpose of the above problem is to emphasize how subtle wording changes affect the problem statement.

- Changing from speed to velocity (velocity could have ± sign while speed is always +).
- "Half the distance" is not the same thing as "half the time".

Check your results with a simulation!

2.15 Jackie throws a ball upwards from the ground with initial speed v. Assume Jackie's height and air resistance are both negligible.

a) Determine the max height.
b) Determine the time to reach max height.
c) Show the time to reach max height is half the total time of flight.
d) Determine the impact velocity when the ball finally returns to the ground.
e) What is the acceleration at max height?

2.16 Lacy and Jimbo stand on the deck of a ship. From height h above the surface of the water, Lacy throws a rock upwards with initial speed v. Jimbo throws a rock downwards from the same height with the same initial speed. The rocks hit the water simultaneously. For both cases assume air resistance is negligible as well as the heights of the people. Jimbo sings karaoke the whole time. Figure to scale (bad attempt at humor). Think: did they throw their rocks at the same instant?

a) Determine impact velocity of Lacy's rock. Answer in terms of g, h, and v.
b) Determine the time of flight of Lacy's rock. Answer in terms of g, h, and v.
c) **Check your results** when $v = 0$ **by comparing them to Thor's hammer** a few problems ago...
d) Determine the time of flight and impact velocity of Jimbo's rock. Answer in terms of g, h, and v.
e) How much earlier did Lacy throw his rock?
f) Determine the total distance traveled by Lacy's rock.

The purpose of this problem is to point out the importance of getting the correct sign on the initial velocity. If interested, run a simulation to compare the two problems. Ophie hangs her head in shame…carry the two of clubs…

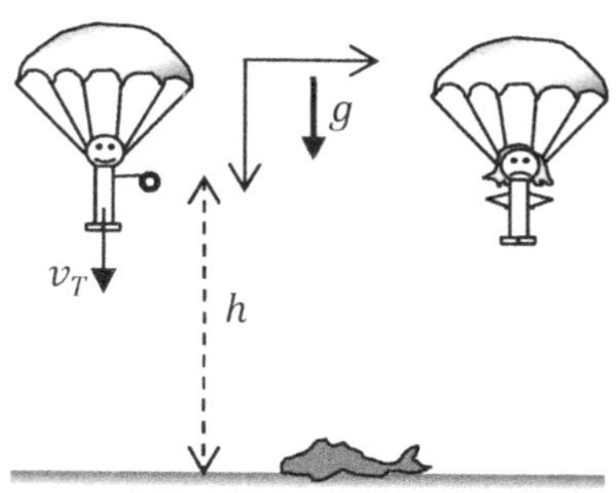

2.17 A parachutist named Captain Xtremo falls from the sky with constant speed v_T. When Captain Xtremo is height h above the ground he drops the ring he plans to use to propose to Debbie. The ring was forged from a magical talisman that cancels the effects of air resistance. The coordinate system is oriented with the positive y-direction *downward.*

a) Determine the speed and velocity of the ring just before impact.
b) Determine the time elapsed between the ring impact and Captain Xtremo reaching the ground.
c) Think: what if h is zero…what should happen to your previous answers? Do your algebraic answers match your reasoning? Assume the height of Xtremo is negligible.

The ring was found by a seal carcass on the beach so it was all good…except now Debbie is stuck with a guy named Captain Xtremo who proposed to her by a seal carcass…

2.18 A projectile is launched vertically with known speed v from an unknown initial height. It is observed that it passes by the launch point after half the flight time has elapsed. Determine the initial height of the building in terms of v and g.

2.19 A giant tortoise doped up with performance enhancing drugs "runs" with constant speed v. A hare jogs 20.0 times as fast as the tortoise if it doesn't try that hard. The tortoise and the hare ran a race measuring 18.9 m somewhere in the Guadalupe dunes. At some point during the race, the hare decided to show off by sitting down and letting the tortoise get ahead of him. After exactly 2.00 min, she resumed her constant rate of travel towards the finish line. Shocking the world, the tortoise won the race by approximately 90.0 cm (about a shell's length).

a) How long was each participant in the race actually moving?
b) What was the average speed of each participant while in motion? Answer in miles per hour.

2.20 A ball is thrown from the ground up to a person standing on a balcony. The balcony is height h above the ground. The ball takes time t to reach the person on the balcony. What is the initial velocity of the ball? Assume air resistance and the heights of the people are negligible.

2.21 A ball is thrown upwards. When the ball reaches ¼ of its maximum height it has speed v. What was the initial speed of the ball?

2.22 Rocket is launched with initial speed of 0 m/s. Initially the thrusters are running and cause a constant upward acceleration with magnitude $2g$. The rocket turns off its thrusters when it reaches a speed of 98 m/s. Once the thrusters are turned off the rocket can be approximated by freefall. What is maximum altitude reached by rocket? If you have already covered plots (next page) you can also plot xt, vt, and at for the rocket.

2.23 A lunar lander in free space can use it's thrusters to accelerate with max magnitude g. The magnitude of freefall acceleration on the moon is roughly $\frac{g}{6}$. When the lander is distance 200 m above the moon it is moving vertically downward with speed $v = 20\ \frac{\text{m}}{\text{s}}$. The lander falls freely until it reaches height h above the surface. From height h until it reaches the ground, the thrusters are fired at the max rate. The speed of the lander is approximately zero just before impact. Assume the amount of fuel burned while the thrusters are on is negligible (not true in real life) so we may treat the system as constant mass. How far above the lunar surface must the astronauts fire the thrusters? Disclaimer: the numbers in this problem are wild guesses, not from research.

2.24 A building has *unknown* height H. Some *unknown* distance above the ground is a window with *known* height h. A ball is released from rest at the edge of the top of the building and passes the window in time t_1. It then takes additional time t_2 to travel from the bottom of the window and impact the ground. Determine the height of the building.

Graphs/plots/charts of Motion
In general, well-made plots can express large amounts of information efficiently. Rather than studying a single data point, entire data sets can be analyzed. Trends in the data can be studied. Comparisons with theoretical equations becomes a visual process requiring mere seconds to internalize.

As a scientist it is expected you can not only create but also interpret plots. Plots of motion are an excellent place to start. In physics we often use time (*t*) on the horizontal axis. We start by focusing on three types of graphs:

- Position versus time
- Velocity versus time
- Acceleration versus time

Notice the vertical axis of each plot represents a vector so $\hat{\imath}$ or $\hat{\jmath}$ is implied on all vertical numbers on the chart.

***xt* graph:**

- Horizontal axis is time (t) while vertical axis is position (x)
- x is position (not distance)
- the *vertical* intercept of the xt plot is the initial position
- the slope at any point on an xt graph is instantaneous velocity ($\frac{d\vec{x}}{dt} = \vec{v}$)
- xt concave up implies acceleration is positive (concave down, acceleration is negative)
- linear xt graph implies constant velocity (positive velocity => positive slope)
- parabolic xt graph implies constant acceleration
- area under xt-graph isn't a useful quantity

***vt* graph:**

- v is *velocity* (not speed)
- the vertical intercept of the vt plot is the initial velocity
- the value of v at any point tells you the *slope* on the xt graph at that same point in time
- the slope of vt curve at any point is instantaneous acceleration ($\frac{d\vec{v}}{dt} = \vec{a}$)
- area under vt curve is **DISPLACEMENT** ($\int_i^f \vec{v}dt = \int_i^f d\vec{x} = \Delta\vec{x}$)
- area under the vt curve is the **CHANGE in position**, not position, not distance!
- linear vt graph implies constant acceleration (positive slope => positive accel)

***at* graph:**

- area under at curve is CHANGE in velocity ($\int_i^f \vec{a}dt = \int_i^f d\vec{v} = \Delta\vec{v}$)
- the value of a at any point tells you the *slope* on the vt graph at that same point in time

Speeding up or slowing down depends on the signs of both acceleration AND velocity.
If they have the same sign the object is speeding up (opposite signs then slowing down).

$v > 0$ AND $a > 0$	moving forward and speeding up
$v > 0$ AND $a < 0$	moving forward and slowing down
$v < 0$ AND $a < 0$	moving backward and speeding up
$v < 0$ AND $a > 0$	moving backward and slowing down

2.25 Plots of a particle's motion are shown at right.

a) Determine the initial position and velocity of the particle.

b) Estimate the slope of the *xt* curve at $t = 3.0$ s using rise over run. Remember to put units on your numbers by reading the axis labels. Compare it to the velocity at 3.0 s.

c) Determine the area under the *vt* curve from 0 to 1.0 s. Keep units on your calculation. This is *displacement*.

d) Use part c (and the initial position $\vec{x}_i = -6\text{ m}\hat{\imath}$) to determine the position at 1.0 s.

e) Determine the area under the *vt* curve from 1.0 to 4.0 s.

f) Use parts c & e to determine the numerical value of total *displacement* after 4.0 sec.

g) Use parts c & e to determine the total *distance* traveled after 4.0 sec.

h) Determine the area under the at curve from 0 to 4.0 s. Use this (and initial velocity) to determine the velocity at 4.0 s. Compare to the vt plot…

i) Determine the slope of the vt curve at 1.0 s. Compare this value to the acceleration at 1.0 s.

j) At what times or time intervals is the object moving right/left/at rest? When is it speeding up and slowing down? Fill in the chart below.

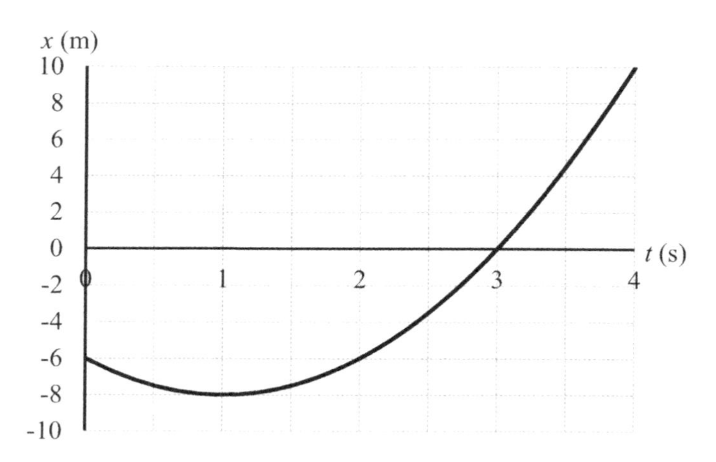

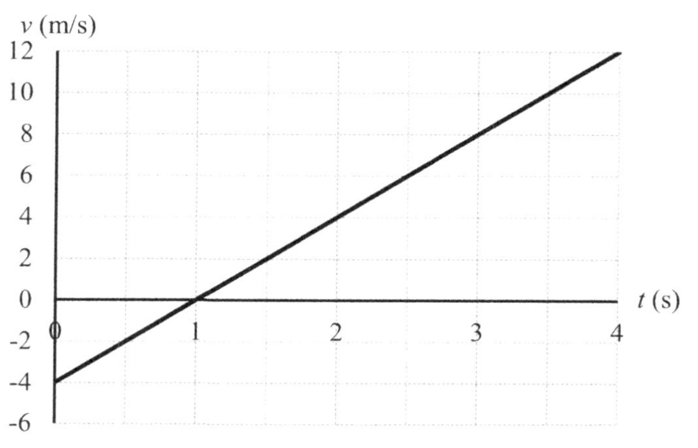

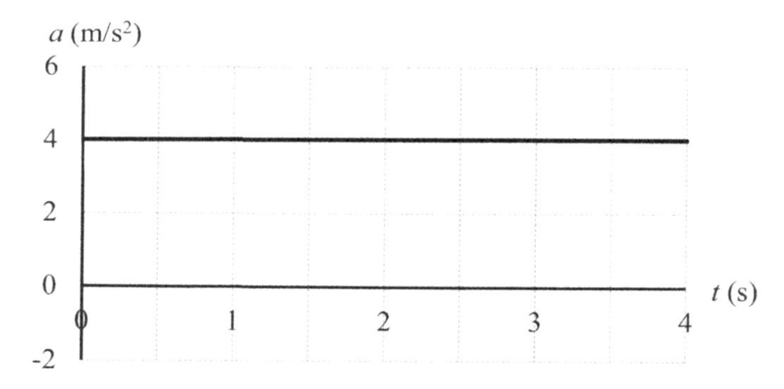

Moving right, speeding up	
Moving right, slowing down	
Moving left, speeding up	
Moving left, slowing down	
At rest	

2.26 Consider checking the plots of problem **2.25** by using a computer simulation…
A suggested simulation is discussed in the supplemental handouts found on robjorstad.com.

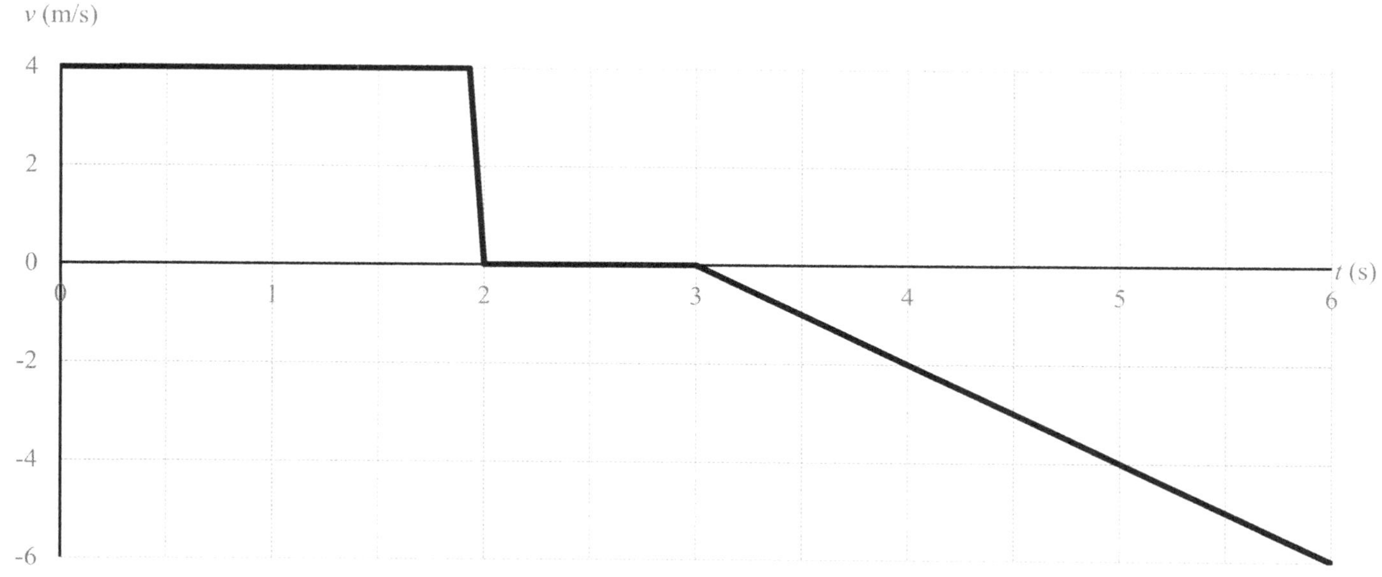

2.27 An object moving in one dimension has the vt-plot is shown above. You are told the object is initially 4.0 m to the left of the origin. While not easily visible, between 1.95 and 2.00 s the object changes speed to 0 m/s. Answer the following questions.

a) When, if at all, is the object speeding up?
b) When, if at all, is the object slowing down?
c) When, if at all, is the object at rest?
d) When, if at all, is the object at constant speed?
e) Determine the displacement on each interval. Also, determine the acceleration over various intervals. Use this information to make xt and at plots. Note: you might need to do a little side problem to get the acceleration between 1.95 and 2.0 s.

Δt (s)	Δx (m)	a (m/s^2)
0 to 1		
1 to 1.95		
1.95 to 2		
2 to 3		
3 to 4		
4 to 5		
5 to 6		

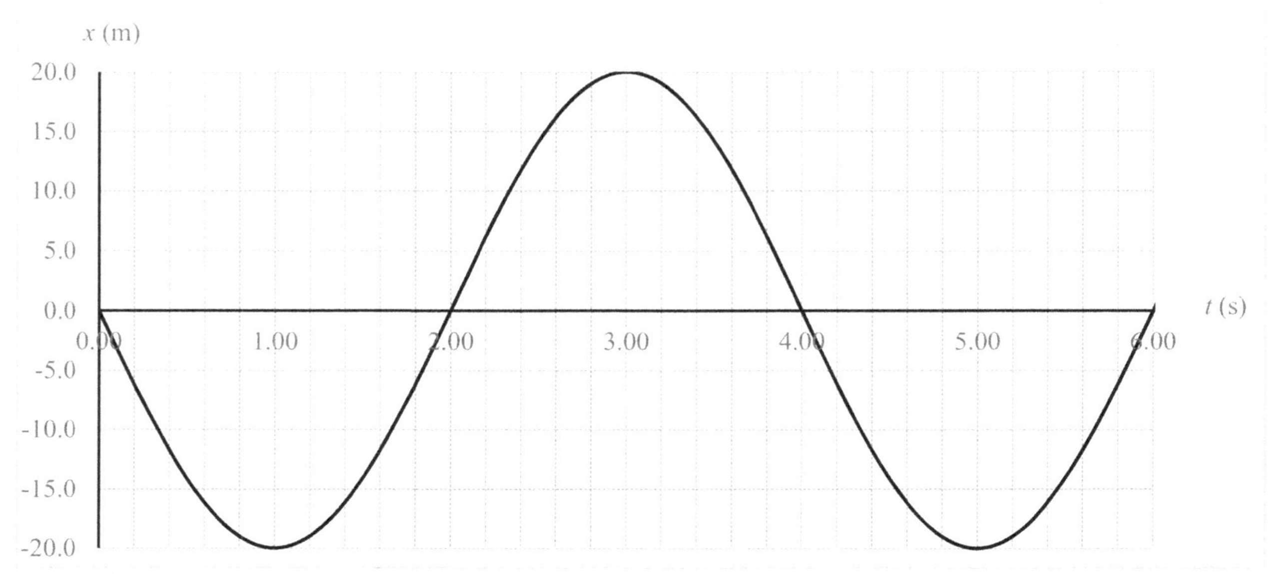

2.28 A plot of *position* versus time for an evil alien named Zykra is shown above. We will assume to the right is the positive direction as usual (so positive numbers imply $+\hat{\imath}$ in the standard way we've done all semester).

a) List time intervals (or instants in time) for which Zykra is moving to the right.
b) Estimate Zykra's velocity at $t = 4.00$ s.
c) List time intervals (or instants in time) Zykra is moving *left* and *slowing down*.
d) List time intervals (or instants in time) Zykra is moving *right* and *speeding up*.
e) Estimate time intervals (or instants in time) for which Zykra's acceleration is negative.

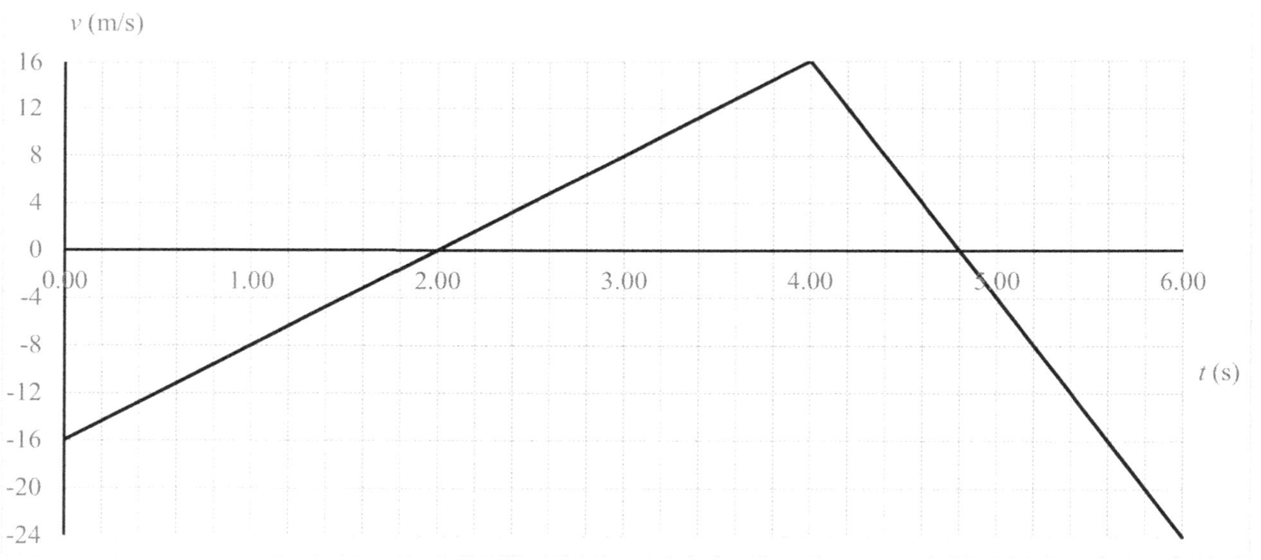

2.29 A plot of *velocity* versus time for an evil alien named Vega is shown above. We will assume to the right is the positive direction as usual (so positive numbers imply $+\hat{\imath}$ in the standard way we've done all semester). Do not assume this graph is linked to the previous as I made up random numbers for both.

a) List time intervals (or instants in time) for which Vega is at rest.
b) Estimate Vega's acceleration at $t = 2.00$ s.
c) List time intervals (or instants in time) Vega is moving *left* and *speeding up*.
d) Estimate the *displacement* between $t = 4.00$ s and $t = 6.00$ s.
e) Determine the *distance* traveled between $t = 4.00$ s and $t = 6.00$ s.
f) Write equations for position and velocity as functions of time for the first 4.00 s of motion.
g) **Challenge:** Write position and velocity as functions of time *after* 4.00 s. This step is a lot trickier…

2.29¼ The graphs of <u>velocity versus time</u> are shown for objects moving in one dimension.

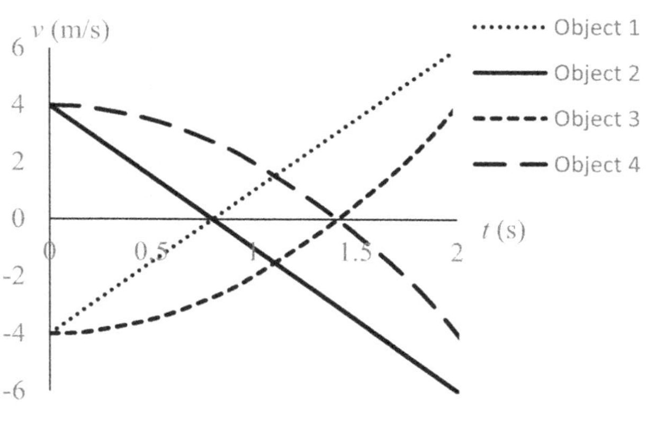

a) Which object or objects have *constant* acceleration? If no object has *constant* acceleration, write "None".

b) Which object or objects have *positive* acceleration? If no object has *positive* acceleration, write "None". To be clear, for this part the acceleration could *changing* but it should always be *positive*.

c) At what time (or over what time interval) is **object 4 at rest**? Assume we are restricting our discussion to times between 0 and 2.00 s. If no such time or interval occurs write "Never happens".

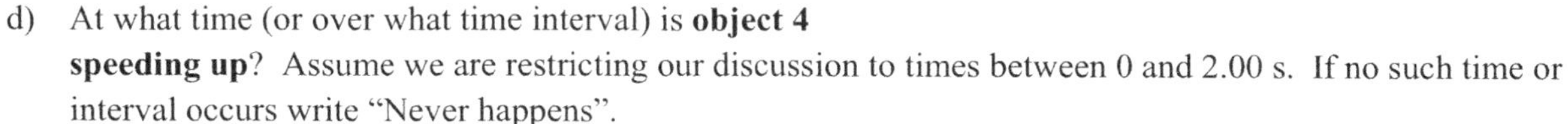

d) At what time (or over what time interval) is **object 4 speeding up**? Assume we are restricting our discussion to times between 0 and 2.00 s. If no such time or interval occurs write "Never happens".

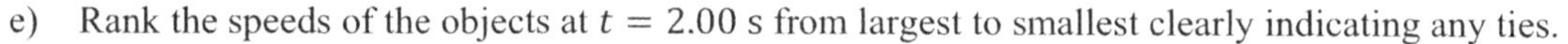

e) Rank the speeds of the objects at $t = 2.00$ s from largest to smallest clearly indicating any ties.

f) For instance, if you think object 1 is the fastest, objects 2 & 3 are tied for second, and object 4 is slowest you should write $v_1 > v_2 = v_3 > v_4$.

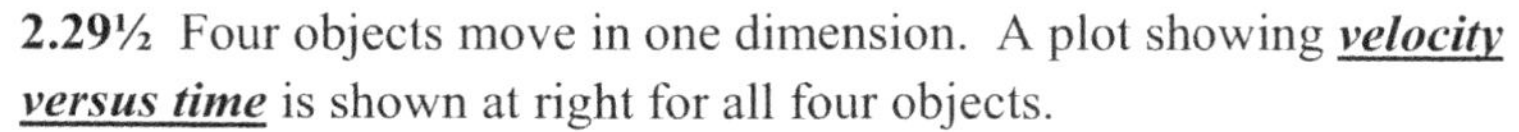

2.29½ Four objects move in one dimension. A plot showing ***<u>velocity versus time</u>*** is shown at right for all four objects.

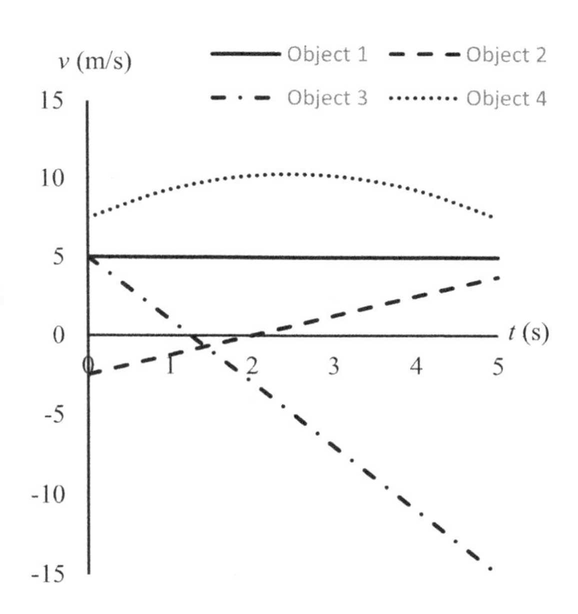

a) Rank the <u>initial</u> *speeds* from smallest to largest clearly indicating any ties.

b) Which objects, if any, have non-zero *constant* acceleration?

c) Which objects, if any, are not *moving*?

d) Which objects, if any, have $a_x = 0$?

e) Which objects, if any, reverse direction?

f) Rank the <u>final</u> *speeds* from smallest to largest clearly indicating any ties.

g) *Monotonically increasing velocity* implies velocity is either constant or increasing. Said another way, $a_x \geq 0$ for all time. Which objects have monotonically increasing velocity?

h) Do any of the objects show monotonically changing *speed*?

i) Rank the *displacements* of each object (most positive to most negative).

j) Rank the *distance traveled* by each object.

2.29¾ Four objects move in one dimension. A plot showing ***position versus time*** is shown at right for all four objects.

a) Rank the initial *speeds* from smallest to largest clearly indicating any ties.
b) Rank the final *velocities* from least negative to most negative.
c) Which objects, if any, have non-zero *constant* acceleration?
d) Which objects, if any, have $a_x \leq 0$ at all points in time (monotonically *decreasing velocity*)?
e) Do any of the objects show monotonically changing *speed*?
f) Which objects, if any, have $a_x = 0$?
g) Which objects, if any, reverse direction?
h) What would the plot of x $vs.$ t look like for an object at rest?
i) Rank the *displacements* of each object (most positive to most negative).
j) Rank the *distance traveled* by each object.

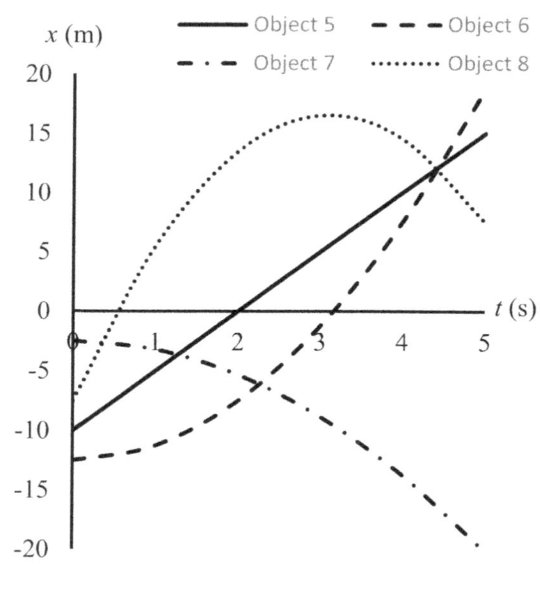

2.29$\frac{7}{8}$ Four objects move in one dimension. A plot showing ***velocity versus time*** is shown at right for all four objects.

a) Rank the initial *speeds* from smallest to largest clearly indicating any ties.
b) Which objects, if any, have non-zero *constant* acceleration?
c) Which objects, if any, are not *moving*?
d) Which objects, if any, have $a_x = 0$?
e) Which objects, if any, reverse direction?
f) Rank the final *speeds* from smallest to largest clearly indicating any ties.
g) Which objects, if any, have $a_x \geq 0$ at all points in time (monotonically *increasing velocity*)?
h) Which objects, if any, have $v_x \geq 0$ at all points in time (monotonically *increasing position*)?
i) Do any of the objects show monotonically changing *speed*?

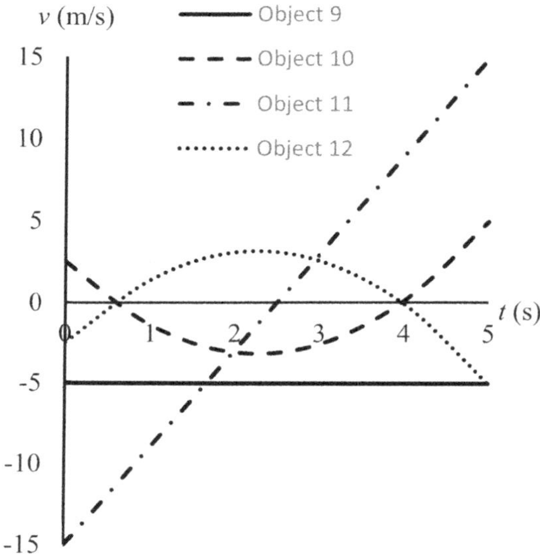

2.29$\frac{15}{16}$ Four objects move in one dimension. A plot showing ***position versus time*** is shown at right for all four objects.

a) Rank the initial *speeds* from smallest to largest clearly indicating any ties.
b) Rank the final *velocities* from least negative to most negative.
c) Which objects, if any, have non-zero *constant* acceleration?
d) Which objects, if any, have $a_x \leq 0$ at all points in time (monotonically decreasing *velocity*)?
e) Which objects, if any, have $v_x \leq 0$ at all points in time (monotonically decreasing *position*)?
f) Which objects, if any, have $a_x = 0$?
g) Which objects, if any, reverse direction?
h) Rank the *displacements* of each object.
i) Rank the *distance traveled* by each object.

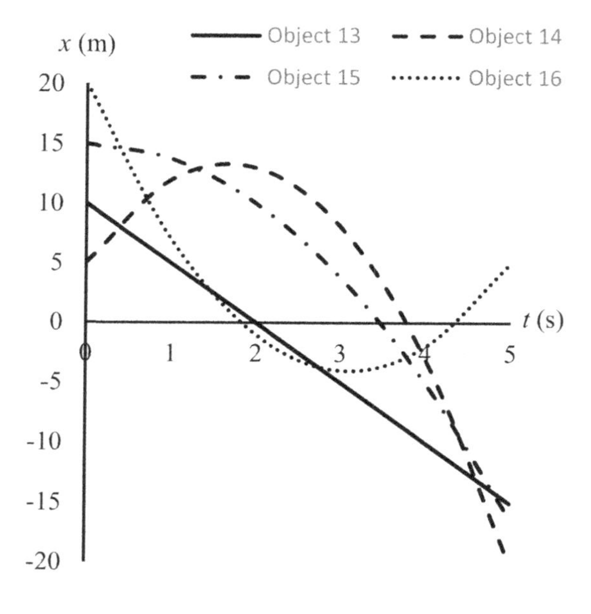

2.30 An object is initially moving *backwards* with speed $2.00\,\frac{\text{m}}{\text{s}}$. After 2.00 seconds has elapsed, the object accelerates uniformly for the next 3.00 seconds. At the end of this first acceleration stage, the object is moving the opposite direction with three times the initial speed. At this point, the object then comes to rest at a constant rate in 1.00 s. Note: in real life an object cannot *instantly* change acceleration. That said, assume the time to change accelerations is negligibly short (too small to be seen on any plot).

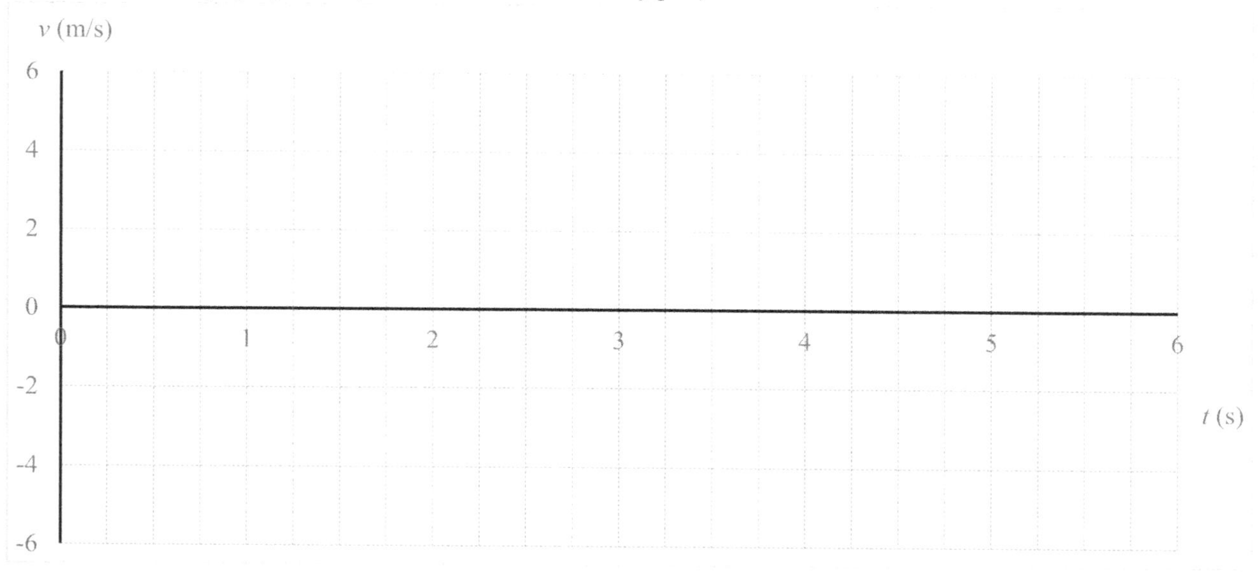

a) Sketch velocity versus time for the object on the plot shown above.
b) At what time, or during what time interval is the *magnitude* of acceleration the greatest?
c) At what time, or during what time interval, is the object at rest?
d) Determine the total *displacement* over the 6.00 second interval. I will assume a negative answer implies displaced to the left.
e) Under what circumstances would the total displacement (answer to previous part) also equal the final position of the object?
f) **Challenge:** At what time, if any, does the object return to the origin?
g) **Challenge:** determine an equation for velocity as a function of time for each time interval.

2.31 Graphs of a vs t are shown for Abigail, Bernie, and Cuba at the bottom of the page. Sketch vt and xt graphs for each person. **Assume each person is initially located 2 m to the right of the origin and is initially moving *to the left* with a speed of 2 m/s.**

Hints: The problem statement above helps you identify the vertical intercept of each plot. Consider the wording carefully to determine if any negative signs are required. Then proceed to construct the velocity graph. Use the fact that acceleration is the slope of the velocity graph. Finally, to make the position graph, use one of two methods:

- use the area under the vt graph to determine displacement (Δx…not x)
- use $x(t) = x_f = x_i + v_{ix}t + \frac{1}{2}a_x t^2$ from constant acceleration kinematics (make a table of t and x values for the first four seconds)

Watch out! It is challenging to correctly use the second method to create position graph for Cuba.

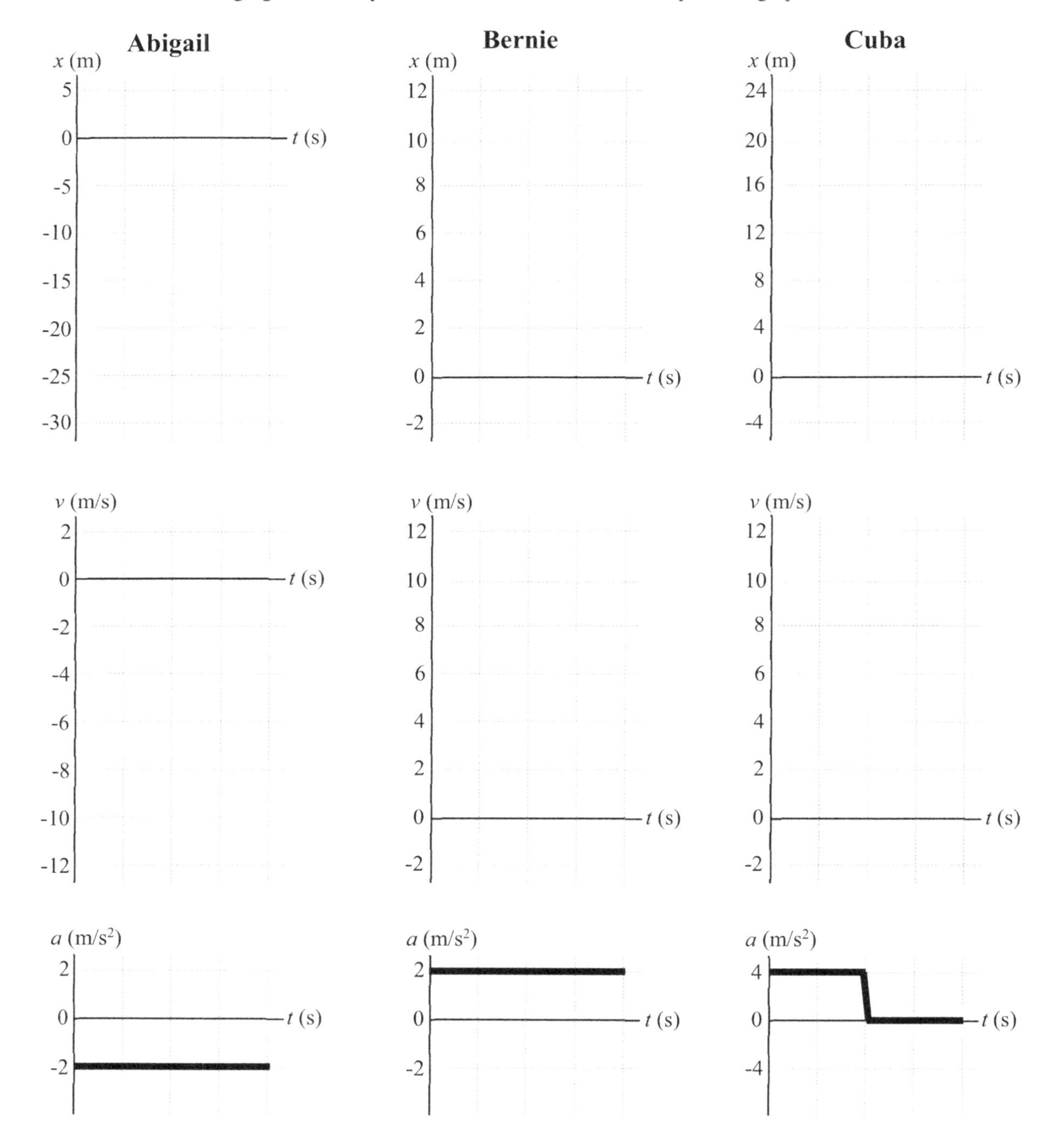

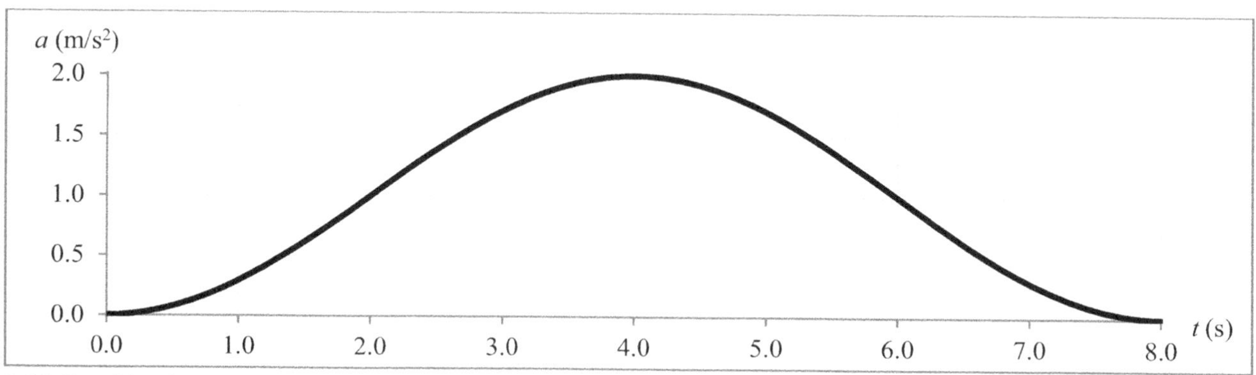

2.32 Consider the a vs t graph shown above. This graph is perhaps more realistic than many of the graphs we have looked at because the acceleration changes gradually. For simplicity, assume the object of interest starts from rest at the origin. This models a car starting from rest at a stop sign which then accelerates.

a) Before doing any math, guess what time the object will have the greatest speed.
b) Estimate the speed of the object at 4.0 sec. Hint: use a triangle and get the area…ignore the curviness.
c) Estimate the speed of the object at 8.0 sec.
d) Describe the slope of the vt graph from 0 to 4.0 sec. Is it + or -? Is it increasing, decreasing, or constant?
e) Describe the slope of the vt graph from 4.0 to 8.0 sec. Is it + or -? Is it increasing, decreasing, or constant?
f) Sketch the vt graph.
g) Consider now the xt graph. When is the concavity of the xt graph be most noticeable? Will it be concave up, concave down, or some mix of these?
h) Are there any times when the xt graph will look essentially like a straight line?
i) Consider the slope of the xt graph. When will the slope be the greatest (near 0, 4 or 8 sec)? Will the slope be positive, negative, flat, or some mix of these?
j) Sketch the xt graph.

2.33 The following questions refer to the motion of an object that has the position versus time graph shown at right. Assume position is measured in m and time in sec. Label time intervals where the object is doing the following:

a) moving right at constant speed
b) moving right with increasing speed
c) moving right with decreasing speed
d) moving left at constant speed
e) moving left with increasing speed
f) moving left with decreasing speed
g) not moving
h) The slope of this curve gives what?
i) The area under this curve gives what? Trick question! Hint: check the units and determine if we use anything with those units.
j) For this question ignore the sharp transitions in the graph at $t = 3, 4, 8$, and 10 sec. That said, for what time intervals the acceleration is positive/negative?
k) Determine the velocity at times $t = 1.5, 3.5$, 8.5, 12.5, 15, 17.5, and 20 seconds.
l) Sketch the vt plot corresponding to this xt plot.

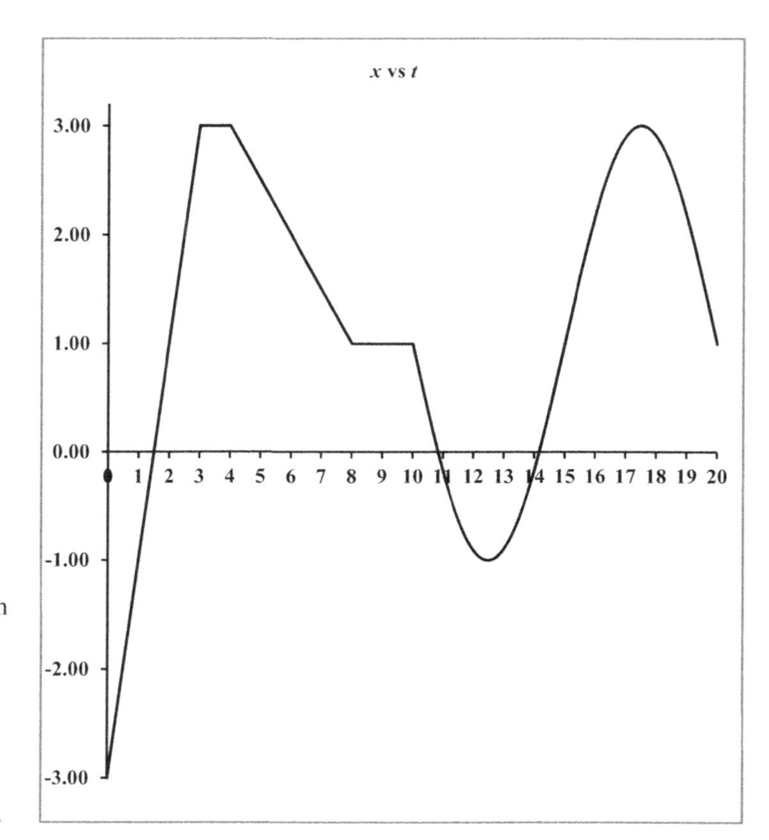

2.34 Look at the following vt graph associated with the 1D motion of a particle. Assume velocity is measured in m/sec and time in sec. You are told the particle is initially located 3.00 m to the left of the origin. List time intervals, if any, where the object is doing the following:

a) moving right at constant speed
b) moving right with increasing speed
c) moving right with decreasing speed
d) moving left at constant speed
e) moving left with increasing speed
f) moving left with decreasing speed
g) not moving
h) The slope of this curve gives what? Hint: check the units if you did rise over run...
i) The area under this curve gives what?
j) Determine the position at times $t = 3, 4, 8$, and 10 seconds. Estimate the position at $t = 12.5$, 15, 17.5, and 20 seconds.
k) Determine the acceleration at times $t = 1.5, 3.5$, 6, and 9 seconds. Estimate a at $t = 12.5$, 15, 17.5, and 20 seconds.
l) Sketch xt & at plots corresponding to this vt plot.

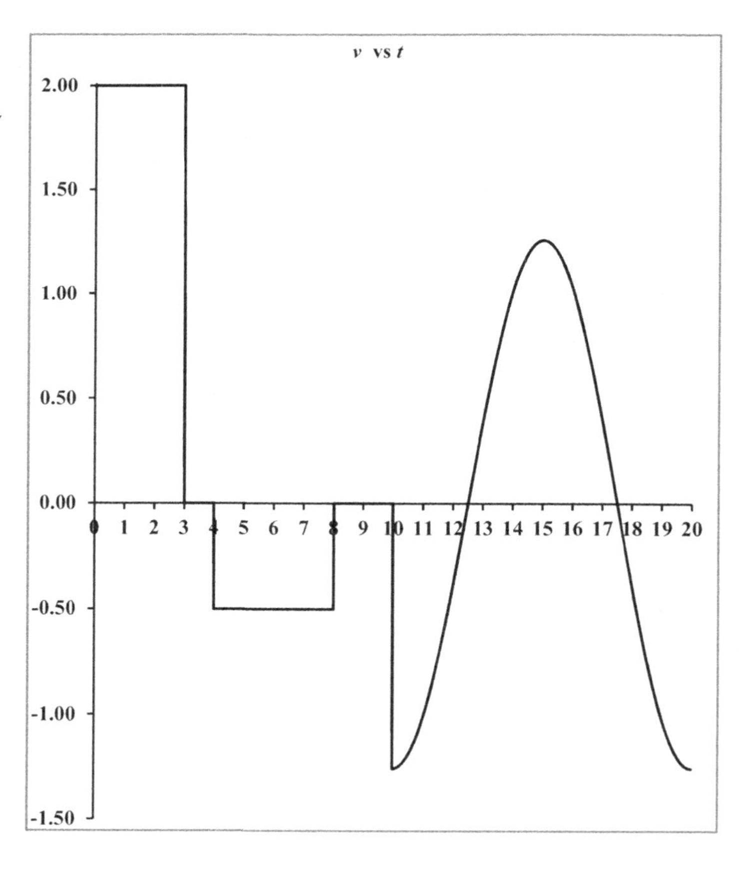

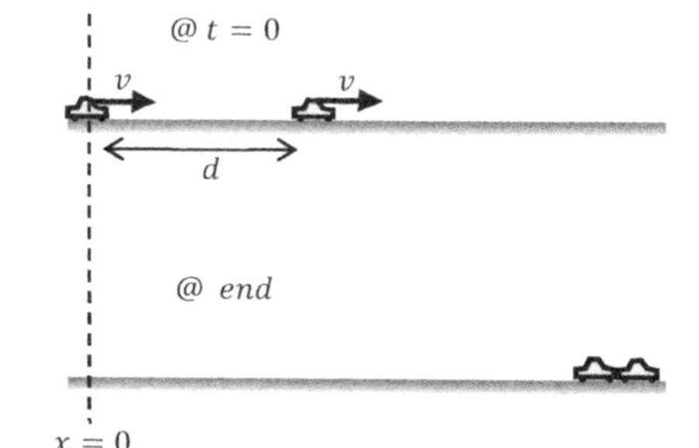

2.35 Two cars are driving in the same direction at the same constant speed v. Initially, the distance from the front bumper of the second car to the rear bumper of the first car is d. At time $t = 0$, the first car gently applies the brakes and slows to a stop with magnitude a. The second car takes a short time t_r to react and then brakes hard, slowing at four times the rate of the first car. As it finally comes to a stop, the second car barely touches the bumper.

a) Determine the distance traveled by each car *during braking*.
b) Determine t_r in terms of d, a, and v?
c) How long is the 1st car in motion in terms of a and v?
d) How long must the 2nd car apply the brakes to stop in terms of a and v?
e) Sketch a plot of v vs t showing both cars. Label the time axis in increments of $\frac{v}{4a}$.
f) Sketch a plot of x vs t showing both cars. Label the time axis in increments of $\frac{v}{4a}$.

2.36 A snail in a car is driving at speed v in a 30 mph zone. He passes a cop without noticing her. At the instant the speeder passes the cop, the cop accelerates from rest with constant rate a. For ease of communication, let us call the point where the speeder passes the cop the origin.

a) What is the maximum distance between the cop and the speeder?
b) How far from the origin is the cop when she finally overtakes the speeder? What is the cop's speed?
c) Think: if the acceleration of the cop is reduced slightly, should the answers to the previous questions increase, decrease, or stay the same? Do your algebraic answers reflect this? What if the speed of the speeder is reduced slightly? What answers would you expect in the extreme limits of $a = 0$ or $v = 0$?
d) Using $v = 20\ \frac{\text{m}}{\text{s}}$ and $a = 5\ \frac{\text{m}}{\text{s}^2}$, plot position versus time for the cop and the speeder on the same graph. Interpret the intersection on this plot.
e) Plot velocity versus time for the cop and the speeder on the same graph. Interpret the meaning of the total area under each curve and the intersection on this plot.

2.37 Suppose a speeder is driving at speed v. He passes a cop without noticing her. At the instant the speeder passes her, the cop accelerates from rest with constant rate a until reaching a speed that is 25% greater than the speeder. The cop continues at this rate until finally catching up to the speeder. For ease of communication, let us call the point where the speeder passes the cop the origin.

a) Determine the maximum distance between the cop and the speeder.
b) Determine the total distance traveled by the cop.
c) Using $v = 20\ \frac{\text{m}}{\text{s}}$ and $a = 5\ \frac{\text{m}}{\text{s}^2}$, plot position versus time for the cop and the speeder on the same graph.
d) Plot velocity versus time for the cop and the speeder on the same graph.

2.38 Suppose a speeder is driving at speed v. She passes a cop without noticing him. At the instant the speeder passes the cop, the cop takes time Δt to put away his equipment before starting to accelerate. The cop accelerates from rest with constant rate a. until reaching the speeder. For ease of communication, let us call the point where the speeder passes the cop the origin.

a) Determine the maximum distance between the cop and the speeder.
b) **Challenge:** Determine the time when the cop reaches the speeder.
c) Use $\Delta t = 2$ s, $v = 20\ \frac{\text{m}}{\text{s}}$ and $a = 5\ \frac{\text{m}}{\text{s}^2}$. Plot position versus time for the cop and the speeder. Use these values to check part b against the intersection of the graph. Numerical methods can be nice...
d) Plot velocity versus time for the cop and the speeder.

This page requires calculus. More practice problems of this type are 2.53-2.55. Those are done algebraically (constants are letters, not numbers).

If position, velocity & acceleration are functions of time we say $x = x(t)$, $v = v(t)$ & $a = a(t)$.
To get $v(t)$ from $x(t)$, take a derivative with respect to time.
To get $a(t)$ from $v(t)$, take a derivative with respect to time.
To get $x(t)$ from $v(t)$, use $x(t) = x_i + \int_0^{t_f} v(t)\, dt$. Notice: in this equation $x(t)$ & x_f are interchangeable.
To get $v(t)$ from $a(t)$, use $v(t) = v_i + \int_0^{t_f} a(t)\, dt$. Notice: in this equation $v(t)$ & a_f are interchangeable.

2.39 Requires Calculus: Initial position of a particle 2.00 m is to the left of the origin. The velocity is given by

$$\vec{v} = (-2.00 + 8.00t - 3.00t^2)\hat{\imath}$$

a) What units are implied on each number in the above equation?
b) Before doing any computations, is it appropriate to use $\Delta x = \frac{1}{2}at^2 + v_0 t$ for this problem? Explain.
c) Compare the initial speed to the initial velocity. Are they the same? Explain.
d) Determine the magnitude of the initial acceleration.
e) Jerk is the time derivative of acceleration ($\vec{J} = \frac{d\vec{a}}{dt}$). Determine the *magnitude* of the initial jerk denoted J_0.
f) Write down the equations for $x(t)$ & $a(t)$. Here I am assuming an implied $\hat{\imath}$ on $x(t)$ & $a(t)$.
g) During what time interval(s), if any, is the object moving to the right?
h) At $t = 3.00$ s, determine the position, speed, and acceleration of the particle?
i) After $t = 3.00$ s, how far has the object traveled?
j) Create plots of xt, vt & at for the first 3 seconds of travel. Use Excel. Use a time increment of 0.1 sec.

2.40 Requires knowledge of derivatives: The position of a piston in an engine is given by $x(t) = -A\cos\omega t$ where $A = 5.00$ cm and$\omega = 209\frac{\text{rad}}{\text{sec}}$. Note: $209\frac{\text{rad}}{\text{sec}} \approx 2000$ RPMs.

a) Determine the initial velocity and acceleration of the piston.
b) Determine the distance traveled by the piston between $t = 0$ and when it first changes direction.
c) Determine the distance traveled by the piston between $t = 0$ and when it first reaches maximum speed.

2.41 Requires knowledge of derivatives: The position of a particle is given by the equation

$$x(t) = \alpha t e^{-\beta t}$$

where α and β are positive constants.

a) Determine the units of α and β.
b) Determine equations for $v(t)$ and $a(t)$.
c) Determine the maximum position of the particle in terms of α and β.
d) Determine the distance traveled (in terms of α and β) after time $t = \frac{2.50}{\beta}$.

2.42 Requires integration and derivatives: A particle initially located 8.00 m to the left of the origin has a velocity given by $v(t) = 16.0t - 2.00t^3$. Note: positive numbers indicate motion in the $\hat{\imath}$ direction.

a) Determine the units assumed on the numbers 2.00 and 16.0.
b) Consider times close to zero. Which term in the velocity equation can be considered negligible? Explain.
c) Consider times far from zero. Which term in the velocity equation can be considered negligible? Explain.
d) Determine the acceleration when the object has reached its maximum positive displacement.
e) If one exists, determine the particle's maximum or minimum speed.
f) Make a sketch of the particle's position versus time.
g) Use Excel to plot position versus time to see how well you did on your sketch.

2.43 Doesn't *require* calculus, but is probably tedious without it: Consider the at-plot for a mecha-walrus shown far below. Assume the mecha-walrus is initially at the origin moving to the *left* with a speed of 2.00 m/s. Try to sketch the xt- and vt-plots. Hint: you can first try to write down an equation for the acceleration as a function of time and then do separation of variables (or just integrate) to get $v(t)$ and $x(t)$. Also, which band name is worse: Mecha-Walrus or Gravel Quarry?

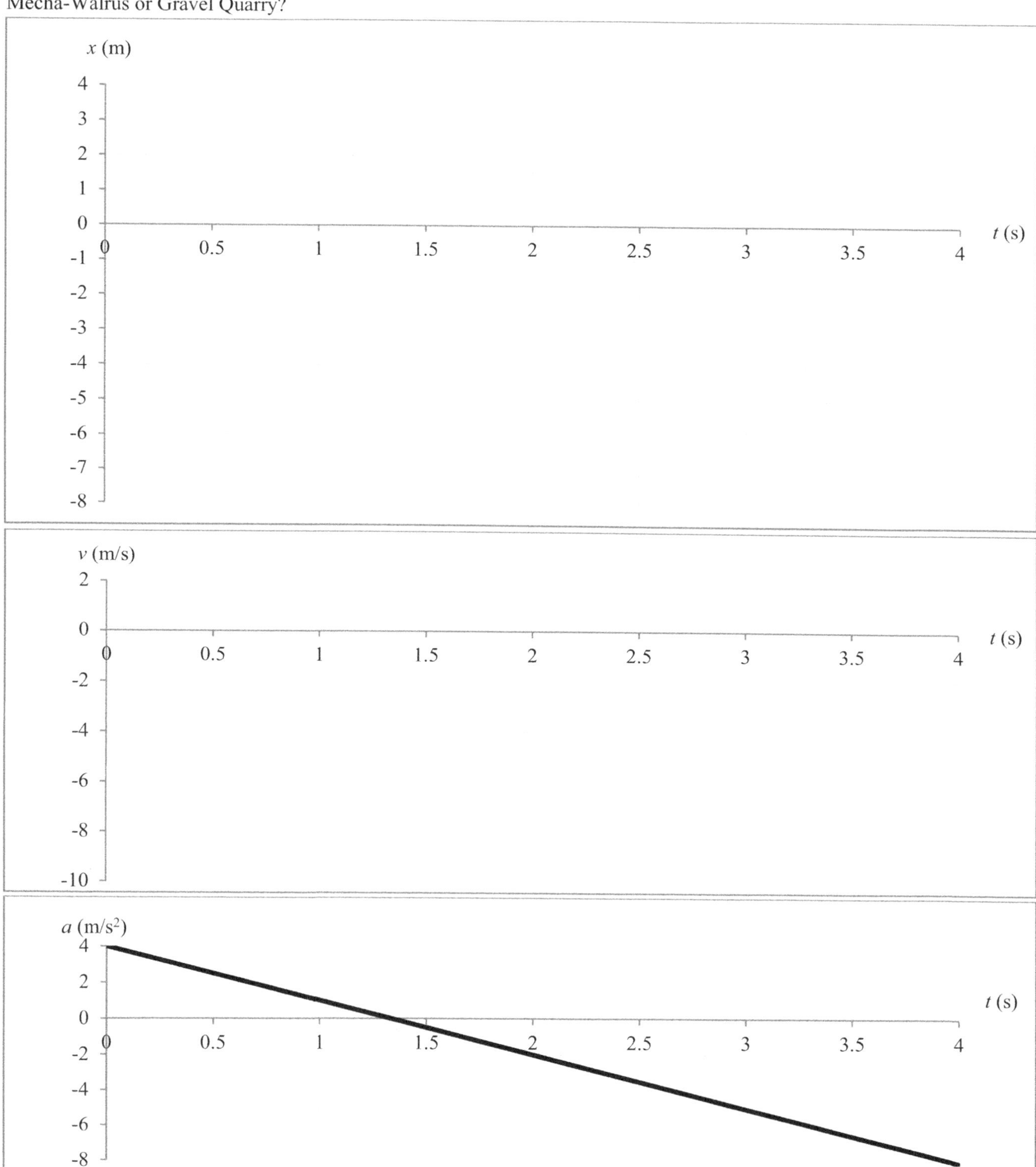

2.44 Look at the following at graph associated with the 1D motion of a particle.

a) What units should you assume apply to each axis?
b) When is the object moving with constant velocity?
c) You are told the object is rest after the 16 second interval shown. What is the initial speed and direction? Answer: 10 m/s to the right.
d) **Challenge:** Sketch a vt plot for this at plot.
e) **Challenge:** get the xt plot as well… you may assume the particle starts from the origin.

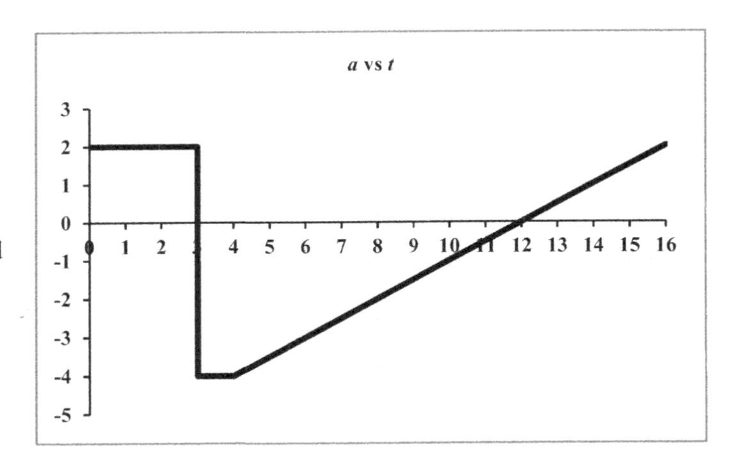

2.45 Answer only (not full solution). A man runs to catch a bus. The man runs towards the stopped bus at with constant rate v. At the instant the man is distance d from the bus, the bus accelerates from rest at rate a.

a) At what time (or times) will he catch the bus?
b) How far does the bus travel in the time required for the man to reach the bus?
c) For what acceleration of the bus would the man not catch it? Hint: the term under the radical must be positive which puts constraints on v, d, and a.

2.46 Consider a race between Carl and Jane. The race is 24 m long. Carl walks half the *distance* at 2 m/s and then runs at 4 m/s the remainder of the race. Jane completes the race by spending the first half of her *time* walking at 2 m/s and the second half of her time running at 4 m/s. Who wins? Is the race a tie? Plot the motion of each runner on the same plot. If someone does finish first, assume they wait at the finish line for the other person to finish.

2.47 A ball is dropped from the roof of a 10.0 m bldg. It hits the ground and bounces straight back up until it reaches a height of only 5.00 m. It then impacts the ground a second time. A person on the roof released a second ball exactly 2.25 seconds after the first ball was dropped from the roof. The *second* ball happens to impact the ground for the *first* time at exactly the same instant the *first* ball impacts the ground for the *second* time. How long was the first ball in contact with the ground? Plot position of each ball on the same graph.

2.48 Not solved yet: A car is driving with initial speed v. In distance d the car slows to half its initial speed. In an additional distance d the car slows to a stop.

a) Before doing anything, take a guess. Which acceleration stage has larger acceleration magnitude…or is it a tie? Which stage take longer…or is it a tie?
b) Determine an algebraic expression for the acceleration in each stage.
c) Determine an algebraic expression for the time of each stage.
d) Without plugging in numbers, sketch what xt, vt & at plots should look like. Then estimate realistic numbers and make the plots in Excel.

2.49 A race requires a contestant to travel from start to finish as shown in the diagram. The diagonal portion is a run through a grassy field at v_1 = 15 km/h while the horizontal portion is a cycle on a straight road at v_2 = 60 km/hr. In this race you are free to start off at any angle you choose as long as you do not cycle in the grass.

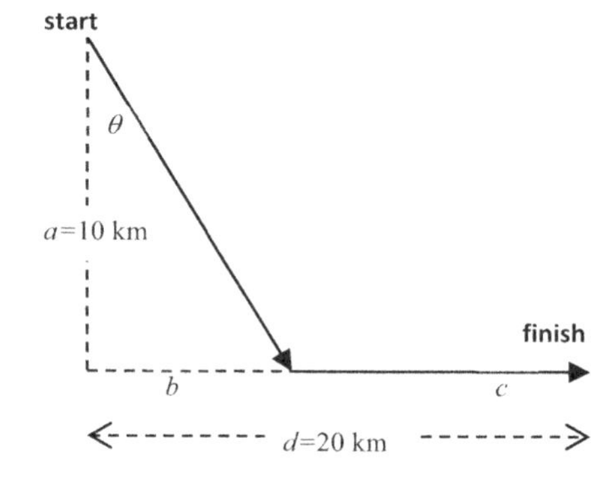

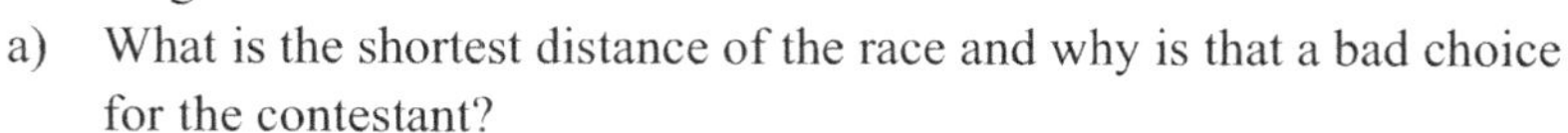

a) What is the shortest distance of the race and why is that a bad choice for the contestant?
b) What choice will get the contestant to the road fastest? Is this a good strategy to minimize the race time?
c) List the givens and unknowns for this problem.
d) For the arbitrary angle θ that is shown, write down a formula for the total race time in terms of the a, b, d, v_1, and v_2. Think, which of these is not constant?
e) **Requires Calculus:** Determine the angle that minimizes the race time. You should find $b = a\frac{v_1}{\sqrt{v_2^2 - v_1^2}}$ and $\theta = 14.5°$.

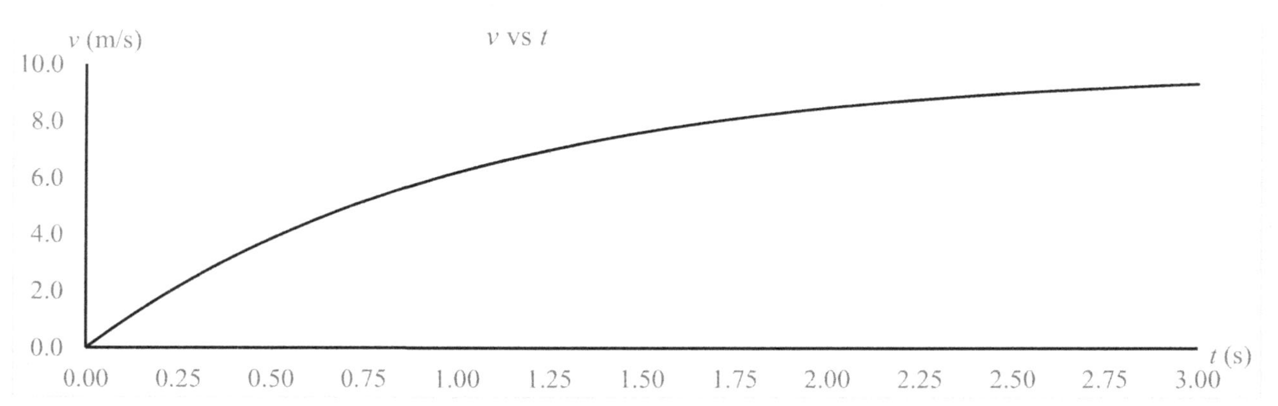

2.50 A marble falling through corn syrup is modeled in problem **2.57**. The marble was released from rest at the origin and we called downwards the positive direction. We assumed $a = g - \frac{b}{m}v$ where m is mass, g is the magnitude of freefall acceleration, and b is a drag constant. To get a feel for the graphs, I used the results of **2.57** to create the *vt* plot shown above. I chose $b = 1\frac{\text{kg}}{\text{s}}$ and $m = 1$ kg.

a) Without thinking too much, whip up a quick sketch of what you think the *xt* and *at* plots look like. Don't worry about numbers at all, just the shape of the graph. Consider this your guess and it is ok to be totally wrong. Then do the following parts to help you improve your guess. Try all parts and then look at the solutions.
b) Estimate the initial acceleration from the graph. Hint: do you use slope or area?
c) Estimate the final acceleration from the graph.
d) Estimate the displacement of the marble between 0 and 0.75 s. Note: when estimating displacements, don't stress over perfect sig figs. Just get a rough estimate as quickly as possible.
e) Estimate the displacement of the marble between 0.75 and 1.5 s.
f) Estimate the displacement between 1.5 and 2.25 s.
g) Estimate the displacement between 2.25 and 3 s.
h) Use your displacements to make a table of position versus time.
i) Use your answers to parts b through d to improve your *xt* and *at* plots. Now look at the answers.
j) On what time interval or intervals is the marble slowing down?
k) **Challenge:** make *at*, *vt*, and *xt* plots using Excel or Matlab. Use the results of problem **2.57** and, for convenience, let $x_i = 0$, $b = 1$ & $m = 1$.

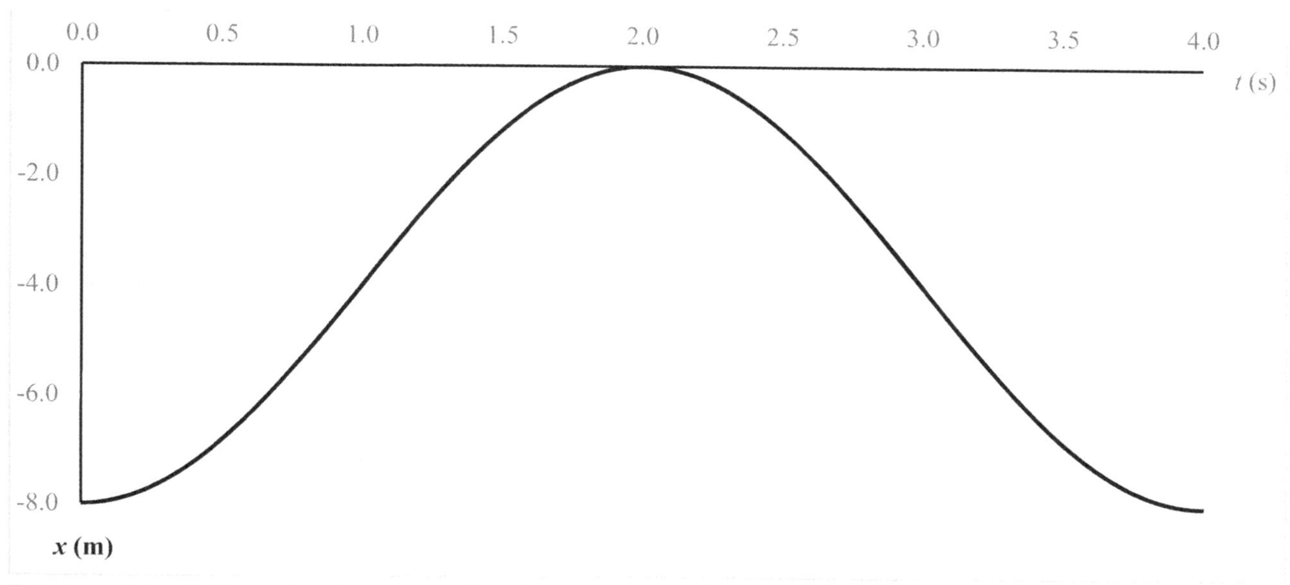

2.51 In problem **2.54** we state the acceleration of a mass on a spring is given by

$$a(t) = a_0 \cos \omega t$$

where the constant a_0 is the acceleration at time $t = 0$ and ω is a constant called the angular frequency. The mass is initially at rest a distance L to the *left* of the origin. A plot of *position* versus time is shown above.

If you are wondering, assume the mass sits on a table with negligible friction. The weight is supported by the table while the mass wiggles back and forth because of the spring.

a) Is the area under the curve more useful for determining the velocity or the acceleration?
b) What is the displacement between 0 and 4 seconds?
c) What is the distance traveled between 0 and 4 seconds?
d) At what times is the mass at rest?
e) When is velocity approximately constant? What does this say about the acceleration?
f) During what time intervals is the object moving to the right and left?
g) At what time (or times) does the object reach max *speed*?
h) Estimate the max speed from the graph.
i) On what time intervals is acceleration positive? When is it negative?
j) When is the object speeding up? When is it slowing down?
k) Sketch plots of vt and at. Don't worry about perfection or getting the numbers correct. Just worry about having the correct sign on v and a at each point in time. Then draw a smooth line connecting the points.
l) **Challenge:** let $L = 8.00$ m, $a_0 = 9.86 \frac{\text{m}}{\text{s}^2}$ and $\omega = 1.57 \frac{\text{rad}}{\text{s}}$. Use the results of problem **2.54** for $x(t)$, $v(t)$ and use them to plot xt, vt, and at from 0 to 4 seconds.

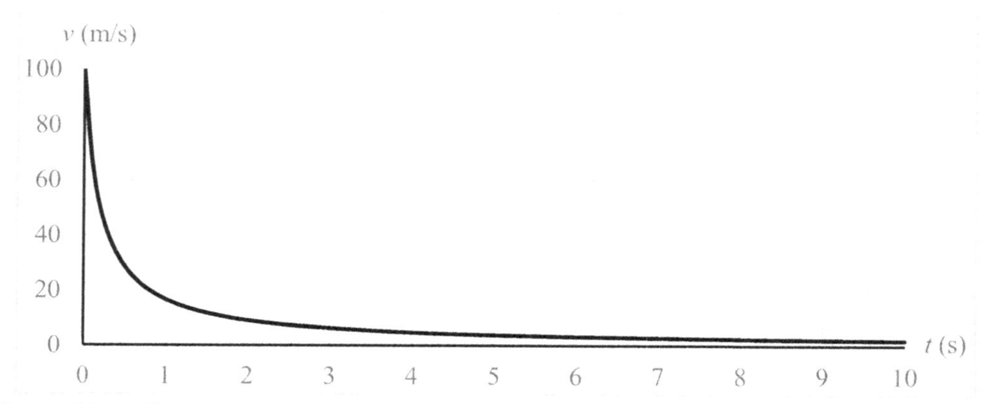

2.52 In problem **2.56** we discuss a rocket-driven sled on a set of rails. When the sled reaches the origin at $t = 0$, the driver turns off the thrusters for the sled. At $t = 0$ the sled has initial velocity $v_0\hat{\imath}$ and acceleration

$$\vec{a} = -\alpha v^2 \hat{\imath}$$

where α is a positive constant. A plot of the velocity versus time is shown using $\alpha = 0.05\ \mathrm{m}^{-1}$ and $v_0 = 100\ \frac{\mathrm{m}}{\mathrm{s}}$.

a) Calculate the magnitude of the initial acceleration in g's. This acceleration would likely kill the driver. Good thing I used robotic driver…

b) Sketch both the xt plot and the at plot. Before looking at the solutions, double check your xt graph for concavity, slope and sign. Double check the slope and sign on your at graph as well.

c) **Challenge:** Use the final results from problem **2.56** to generate plots of xt, vt, and at. **Warning:** in this problem the acceleration is a function of *velocity*, not a function of *time*. **Separation of variables is required. Separation of variables is discussed on the ensuing pages...**

Non-Constant Acceleration

You might be wondering about situations like the previous graph where acceleration is not-constant. We can't even split the graph into pieces where acceleration is constant. Differential equations describe the motion. The solutions to the differential equations provide the equations of motion. Let's start with a familiar example to learn the names of all the pieces

Differential equation	$dv = adt$ where a is constant
Solution to Differential Equation	$v(t) = v_i + at$
Differential velocity	dv
Differential time	dt

Notice the diff eq is simply another way of writing $a = \frac{dv}{dt}$. Anything with a *"d"* in front of it is called a differential.

We will learn one technique, called separation of variables, to solve differential equations for non-constant acceleration. This technique is limited but it is enough to help us study things like drag forces, the charging of capacitors, the energizing of magnetic solenoids, radioactive decay, & the Schrödinger model of the hydrogen atom.

Systems of differential equations come into play in predator-prey (Lotka-Volterra) models, tsunami models, and many other applications. The Navier-Stokes equations look both interesting and gnarly. Many times differential equations are solved with computer programs. Take a course on numerical ODEs or something like that…

Here is an example of how I like to use separation of variables for a random diff eq:

We are given the following acceleration	$a = \frac{6}{v}t^2$
Rewrite using differentials	$\frac{dv}{dt} = \frac{6}{v}t^2$
Separate the variables Get v's with v's & t's with t's	$vdv = 6t^2dt$
Integrate both sides including limits	$\int_i^f vdv = \int_i^f 6t^2dt$
Doing the integration	$\left[\frac{v^2}{2}\right]_i^f = [2t^3]_i^f$
Putting in limits	$\frac{v_f^2}{2} - \frac{v_i^2}{2} = 2t_f^3 - 2t_i^3$
AFTER INTEGRATION Multiply all terms by 2 then set $t_f = t$ & $t_i = 0$	$v_f^2 - v_i^2 = 4t^3$
Solve for v_f	$v_f = \pm\sqrt{v_i^2 + 4t^3}$
Interpret v_f as $v(t)$ think about $\pm$ sign based on particular problem	$v(t) = \pm\sqrt{v_i^2 + 4t^3}$

Comments:

- Most problems typically start at time $t = 0$ with a known v_i. By putting in the limits we are essentially saving a step in determining the constant C associated with indefinite integrals.
- $\int_i^f dv = [v]_i^f = v_f - v_i = \Delta v \neq v$
- The previous bullet is the same as remembering $\int dv \neq v$ but rather $\int dv = v + C$
- Strictly speaking we shouldn't have the same variable t in the integrand and the limit. This is why I wait until after the integration to set $t_f \rightarrow t$. Unless you are a hard core mathlete, you probably don't care about this and can never think about it again.
- When a is a polynomial in t you can simply integrate to get v like you probably did in calc class.
- **When a depends on v, step 3 cannot be skipped.**

Example of most common student MISTAKE in Separation of Variables

In one air resistance model $a = -\alpha v^2$ where α is a constant. Let's do the process and I'll point out where the common mistake is made.

We are given the following acceleration	$a = -\alpha v^2$
Rewrite using differentials	$\frac{dv}{dt} = -\alpha v^2$
Multiply by dt	$dv = -\alpha v^2 dt$
Integrate both sides including limits	$\int_i^f dv = \int_i^f -\alpha v^2 dt$
Doing the integration	$[v]_i^f = [-\alpha v^2 t]_i^f$ **BAD MONKEY**

Everything is fine, in theory, until Step 5. **Think about it:** the object is accelerating so v cannot be constant in time. How could it factor out of the integral as a constant!?!?!?!? Do not be a bad monkey...

To fix the mistake, do it like this instead:

We are given the following acceleration	$a = -\alpha v^2$
Rewrite using differentials	$\frac{dv}{dt} = -\alpha v^2$
Separate the variables Get v's with v's & t's with t's	$\frac{dv}{v^2} = -\alpha dt$
Integrate both sides including limits	$\int_i^f \frac{dv}{v^2} = \int_i^f -\alpha dt$

Notice this will give a wildly different answer than the incorrect process at the top of the page. In case you are wondering, the constant $-\alpha$ can go on either side but most tend to group constants with the t's. In the end you solve for $v_f = v(t)$ so this saves a step.

The following are examples of differential equations used in your the first few physics courses.

Practice problems start on the next page.

Some of these will appear over there...

1) $a = \frac{dv}{dt}$

2) $v = \frac{dx}{dt}$

3) $-bv = m\frac{dv}{dt}$

4) $mg - bv = m\frac{dv}{dt}$

5) $-bv^2 = m\frac{dv}{dt}$

6) $mg - bv^2 = m\frac{dv}{dt}$

7) $-kx = m\frac{d^2x}{dt^2}$

8) $-kx - b\frac{dx}{dt} = m\frac{d^2x}{dt^2}$

9) $F\sin\omega t - kx - b\frac{dx}{dt} = m\frac{d^2x}{dt^2}$

10) $Mdv = -v_e dM$

The problems on this page do not require you to separate before you integrate.
The problems on the next page do require separation prior to integration.

2.53 Jerk (J) is the time derivative of acceleration given by

$$\frac{da}{dt} = J$$

Let us assume a particle has constant jerk. At time $t = 0$, the particle has initial positon x_0, initial velocity v_0, and initial acceleration a_0. Note: the units of J are $\frac{m}{s^3}$

a) Determine $a(t)$ using separation of variables starting with $\frac{da}{dt} = J$ where J is constant.
b) Determine $v(t)$. Hint: do separation of variables again using $a = \frac{dv}{dt}$. Substitute in your result for $a(t)$.
c) Determine $x(t)$. Hint: use $v = \frac{dx}{dt}$. Substitute in your result for $v(t) = v$.
d) Consider your results in the case of $J = 0$. Is acceleration constant? Do you get the constant acceleration equations of motion?

2.54 The acceleration of a mass on a spring is given by

$$a(t) = a_0 \cos \omega t$$

where the constant a_0 is the acceleration at time $t = 0$ and ω is a constant called the angular frequency. Angular frequency has units of $\frac{\text{rad}}{\text{s}}$ (similar to RPMs). We will learn more about angular frequency in Chapter 10 on rotational motion. The mass is initially at rest a distance L to the *left* of the origin. In case you are wondering, how this mass-spring system can work while *horizontal* (instead of vertical), assume the mass sits on a table with negligible friction. The weight is supported by the table while the mass wiggles back and forth.

a) Determine the velocity as a function of time. Remember to check the units of your results.
b) Determine the position as a function of time.
c) Determine the jerk as a function of time.

2.55 The velocity of a particle is given by the equation

$$v(t) = \alpha t e^{-\beta t}$$

where α and β are positive constants. Assume the particle is initially located at the origin.

a) Determine the units of α and β.
b) Determine equations for $a(t)$ and the jerk $J(t) = \frac{da}{dt}$.
c) Determine an equation for $x(t)$.
d) Does the particle reach a max position? Explain.

2.56 Requires separation before integration: A rocket-driven sled is on a set of rails constructed in the desert. When the sled reaches the origin at $t = 0$, the driver turns off the thrusters. The sled has initial velocity $v_0\hat{\imath}$ and acceleration given by

$$\vec{a} = -\alpha v^2 \hat{\imath}$$

where α is a positive constant. Note: here acceleration is a function of *velocity* (not a function of *time*).

a) Determine the units on α.
b) Determine the velocity as a function of time.
c) **Challenge:** determine $x(t)$.

2.57 Requires separation before integration: For a marble released from rest dropping through corn syrup we can model acceleration as

$$a = g - \frac{b}{m}v$$

where m is mass (constant), g is the magnitude of freefall acceleration (constant), and b is a drag constant. This equation assumes down is the positive direction for all vectors. Note: this *may not* be the *best* model, but it is better than no model. Note: here acceleration is a function of *velocity* (not a function of *time*).

a) Determine the units of b.
b) Determine $v(t)$.
c) Determine $a(t)$.
d) **Challenge:** determine $x(t)$.

2.58 Requires separation of variables: A skydiver jumps from a stationary helicopter distance h above the ground. We can approximate the skydiver as being initially at rest. The acceleration as a function of time is

$$a = g - \alpha v^2$$

where α is a positive constant. Note: here acceleration is a function of *velocity* (not a function of *time*).

a) Determine the units of α.
b) **Challenge:** determine the velocity as a function of time.
c) **Challenge:** determine $x(t)$.

It is interesting to note 4 distinct cases of high velocity drag. A ball can be thrown upwards or downwards. In each case the initial velocity can be either above terminal velocity or below terminal velocity. The signs of $\vec{v}$ and the size of v (relative to terminal velocity) force one to choose different solutions to the integrals that appear when doing separation of variables. Sometimes you get tangent function, other times hyperbolic tangent. For upwards throws above terminal velocity, the equation modeling the ball's motion will switch at terminal velocity (which also happens to be the equilibrium velocity). Then the equation will switch yet again when the ball reverses direction at max height!!! I don't know why I find this so interesting but I do. Maybe it seems neat to you as well.

You know what else is neat? Making a code to simulate the flight and never actually needing any of the three analytic solutions at all!!!!

While I'm at it, what is even MORE neat? Making a three dimensional computer code to handle the flight. Then, use the analytic solutions in 1D, found in the problem above, to check your computer code in 1D before trusting it to work properly in 3D. Think: we all make typos in the real world. If you can analyze a problem in multiple ways (with both analytic AND simulation methods) you can use each result as a double check on the other and gain deeper understanding. Finally, compare your simulation and analytic solutions to video capture of actual balls being thrown around and see if your predictions match experiment for a wide range of initial conditions. This is essentially the scientific method. These cheesy little book problems do have a place in the grand scheme of things, even if it doesn't seem like it now.

Vectors, Vector Components, Scalar Components

A vector has *magnitude* and *direction*.
An example is velocity which has magnitude (speed) and direction (heading).

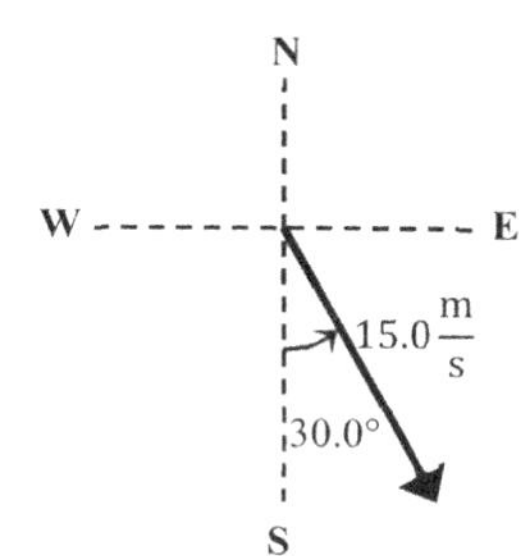

Suppose we have a velocity that is 15.0 m/s with heading 30.0° east of south. This vector is shown in the figure at right. The cardinal directions (north, south, east, & west) are shown as they are typically drawn with **N** as the positive y-axis and **E** as the positive x-axis.

The standard conventions for describing the vector, vector components, and scalar components are shown below. Please pay particular attention to when the $\hat{\imath}$ and $\hat{\jmath}$ are included. Also notice, "east of south" implies you first align with south and then go east from it.

Vector (<u>Polar</u> Form)	$\vec{v} = v$ @ some angle	**Vector (<u>Cartesian</u> Form)**	$\vec{v} = v_x\hat{\imath} + v_y\hat{\jmath}$
Vector (Polar Form 1)	$\vec{v} = 15.0\frac{\text{m}}{\text{s}}$ @ 30.0° E of S	**Vector (Cartesian Form)**	$\vec{v} = 7.50\frac{\text{m}}{\text{s}}\hat{\imath} - 13.0\frac{\text{m}}{\text{s}}\hat{\jmath}$
Vector (Polar Form 2)	$\vec{v} = 15.0\frac{\text{m}}{\text{s}}$ @ 300.0°	**Vector Components**	$\vec{v_x} = 7.50\frac{\text{m}}{\text{s}}\hat{\imath}$ & $\vec{v_y} = -13.0\frac{\text{m}}{\text{s}}\hat{\jmath}$
Vector (Polar Form 3)	$\vec{v} = 15.0\frac{\text{m}}{\text{s}}$ @ -60.0°	**Scalar Components**	$v_x = 7.50\frac{\text{m}}{\text{s}}$ & $v_y = -13.0\frac{\text{m}}{\text{s}}$

The following are several points worth stressing:

- In polar form 1, my convention is often to reference the closest cardinal direction.
- In polar forms 2 & 3, by convention angles are referenced to the positive x-axis.
- Vector and scalar components both have ± signs
- Include the arrow ($\overrightarrow{\ }$) the $\hat{\imath}$ and $\hat{\jmath}$ on the *vector* components but not on the *scalar* components
- Do not say "-60° below" the x-axis. Just as "-2 to the left" implies "+2 to the right", "-60° *below* the x-axis" implies "+60° *above* the x-axis".

Some other useful vector information

The symbol indicating the *magnitude* of a vector is written in one of two ways:

$$\textit{magnitude} \text{ of vector } \vec{F} = \|\vec{F}\| = F = \sqrt{F_x^2 + F_y^2}$$

In 2D, one can determine the direction using SOH CAH TOA (see figure at right).

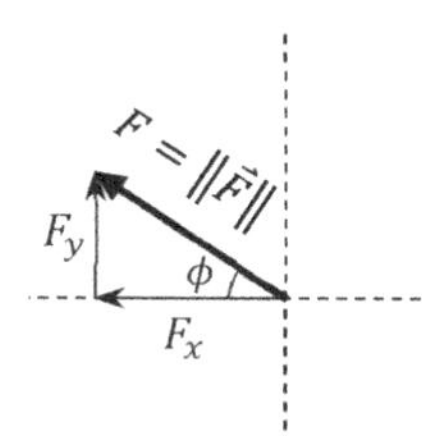

$$\phi = \tan^{-1}\left|\frac{F_y}{F_x}\right|$$

Note: I use the absolute value then draw pictures to ensure I understand the direction.
I feel doing this process forces me to check & double check my own understanding.
I let the arrowheads in my picture remind me which terms are positive or negative.
This may seem strange now, but pretty much all physicists do this all the time in chapters 5 & 6 (forces).

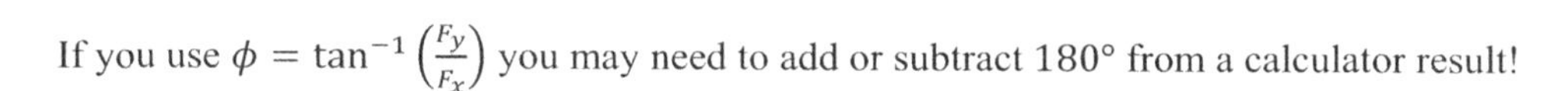
If you use $\phi = \tan^{-1}\left(\frac{F_y}{F_x}\right)$ you may need to add or subtract 180° from a calculator result!

In 3D, the *magnitude* equation becomes $F = \sqrt{F_x^2 + F_y^2 + F_z^2}$.

In 3D, the *direction* is best described by a unit vector given by $\hat{F} = \frac{\vec{F}}{F}$. More on that soon…

3.1 Consider the following vectors. Write down each vector in Cartesian form.

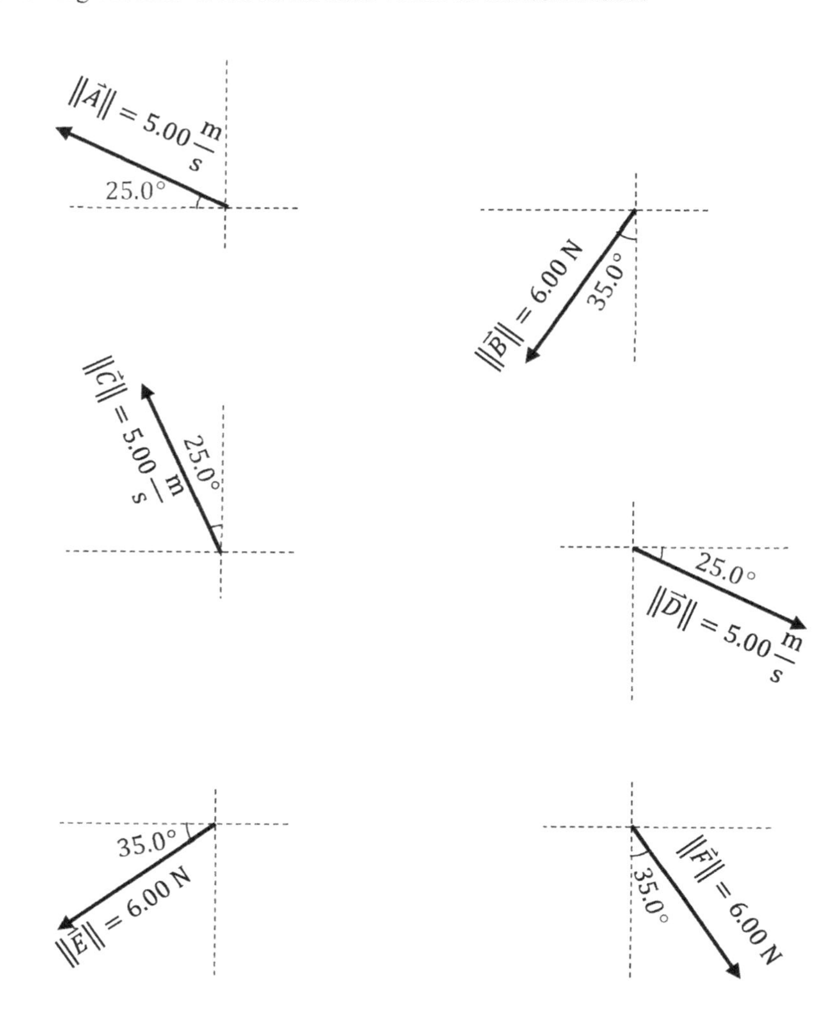

3.2 Determine the magnitude and direction of each vector. To clearly label the direction, sketch a picture and label the one of the angles.

$\vec{G} = (-12.00\hat{\imath} - 5.00\hat{\jmath})\text{m}$	$\vec{H} = (-4.00\hat{\imath} + 3.00\hat{\jmath})\text{m}$

3.3 Switching from magnitude and direction (polar form) to unit-vector notation (Cartesian form).
It is easy to mix up ± signs or switch sine and cosine when determining components. It is worth it to verify you are doing this correctly before moving on to performing mathematical operations with vectors. Write each vector using unit-vector notation. Some answers turn out the same. After computing components, check the signs and sizes of each component with a sketch.

$\vec{A} = 30.0\frac{\text{m}}{\text{s}}$ directed 150° from the positive x-axis	$\vec{D} = 30.0\frac{\text{m}}{\text{s}}$ directed 30.0° below the negative x-axis
$\vec{B} = 30.0\frac{\text{m}}{\text{s}}$ directed -30.0° from the positive x-axis	$\vec{E} = 30.0\frac{\text{m}}{\text{s}}$ directed 30.0° west of north
$\vec{C} = 30.0\frac{\text{m}}{\text{s}}$ directed 30.0° above the negative x-axis	$\vec{F} = 30.0\frac{\text{m}}{\text{s}}$ directed 30.0° north of west

Note to teachers: in most vector addition solutions I used green for $\vec{A}$, blue for $\vec{B}$, red for $\vec{C}$, and black for $\vec{R}$.

Component-wise Vector Addition

When adding vectors component-wise (using $\hat{\imath}$ and $\hat{\jmath}$), use the following procedure

Example is worked on next two pages (shown in two different styles).

1) Split the vectors into components using trig and/or geometry.
2) Double check the signs and relative sizes
   a. Did you flip a sine with a cosine by accident?
   b. Did you forget to put in a minus sign?
3) Add the components together to obtain the resultant vector
4) Convert the components to polar form (magnitude and direction)

Note: the order in which you add vectors is unimportant.

Graphical Vector Addition

To add vectors graphically, follow this procedure.

1) Place the tail of the first vector at the origin.
2) At the tip of the first vector, sketch a tiny new coordinate system. The positive y-direction still points straight up.
3) Put the tail of the second vector at the origin of this new coordinate system.
4) Repeat until all vectors are used.
5) The resultant is found by connecting the first tail to the last tip.

3.4 Assume the spacing between adjacent gridlines is 10.0 $\frac{\text{m}}{\text{s}}$.

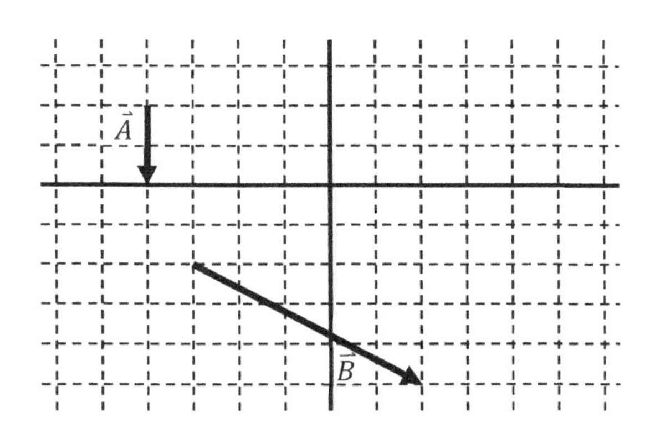

Here the vectors $\vec{A}$ and $\vec{B}$ represent velocity vectors with a magnitude (speed) and direction (heading). Manipulation of these vectors would be important to perhaps an airplane pilot. One might be wind velocity. Another might be the plane velocity.

a) Write down each vector in Cartesian component form.
b) Write down each vector in polar form 1 (angle to the nearest cardinal direction).
c) A third vector $\vec{C}$ has speed 50.0 $\frac{\text{m}}{\text{s}}$ with heading 36.87° west of north. Determine this vector in Cartesian component form and draw it somewhere on the upper grid. Recall the tail need not be at the origin.
d) Find a partner. Have each person choose a different order for the vectors and do graphical addition method (tail-to-tip) to add your three vectors on the fresh grid at right. Verify you get the same resultant.
e) Verify you obtain the same result using component wise addition.

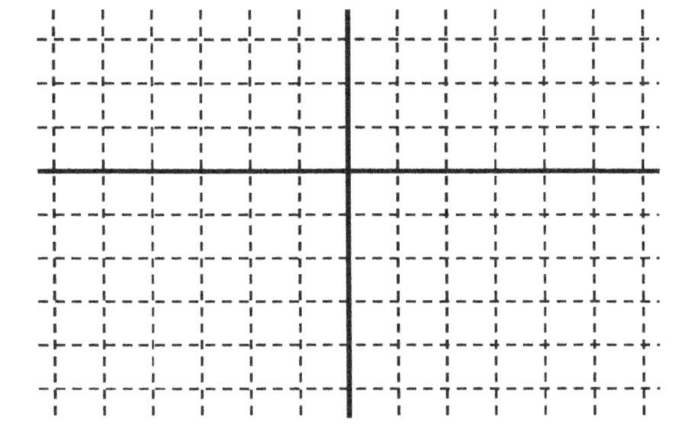

Note: for this special case graphical addition was probably easier. In general component-wise addition is faster and more precise.

Example of vector addition

Two vectors are given as $\vec{A} = 5.00$ @ $53.1°$ W of N and $\vec{B} = 10.00$ @ $36.9°$ S of W.
Determine $\vec{R} = \vec{A} + \vec{B}$.

Procedure

1) Sketch each vector to get a feel for where they point and label the angles.
2) Rewrite in Cartesian form.
   a. Double check the signs of each component of each vector.
   b. Based on your sketch, verify you appropriately used sine/cosine for each component.
3) Add $\hat{\imath}$'s to $\hat{\imath}$'s and $\hat{\jmath}$'s to $\hat{\jmath}$'s.
4) Sketch the final vector $\vec{R}$.
5) Use your sketch to determine $\vec{R}$ in polar form.
   a. Use the Pythagorean theorem to determine the magnitude.
   b. Use SOH CAH TOA to get the angle (usually $\tan^{-1}$).
6) Verify the angle is correct using graphical addition (tail-to-tip method).

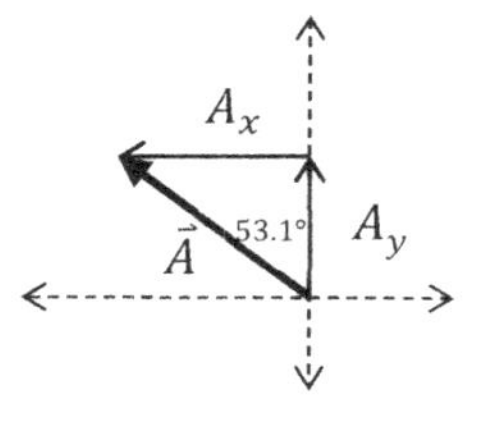

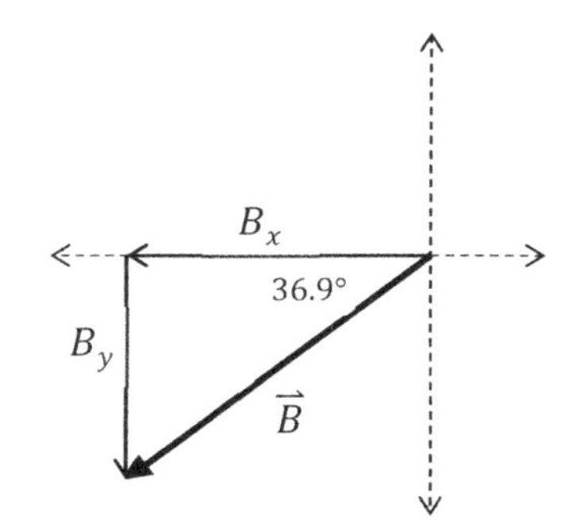

$\vec{A} = A_x\hat{\imath} + A_y\hat{\jmath}$
where $A_x = -5.00 \sin 53.1°$ and $A_y = +5.00 \cos 53.1°$.
The ±'s come from the arrowheads in sketch. Notice that cosine and sine are switched from the standard math convention! In standard math notation the angle is always to the horizontal.

$\vec{B} = B_x\hat{\imath} + B_y\hat{\jmath}$
where $B_x = -10.0 \cos 36.9°$ and $B_y = -10.0 \sin 36.9°$.
The ±'s come from the arrowheads in sketch.

Now we see

$$\begin{aligned} \vec{A} &= \quad -5.00 \sin 53.1°\,\hat{\imath} + \ \ 5.00 \cos 53.1°\,\hat{\jmath} \\ \underline{+\vec{B}} &\underline{= -10.0 \cos 36.9°\,\hat{\imath} - 10.0 \sin 36.9°\,\hat{\jmath}} \\ \vec{R} &= \quad -11.9\underline{9}5\,\hat{\imath} \quad + \quad -3.0\underline{0}2\hat{\jmath} \end{aligned}$$

Notice the extra sig fig on the $\hat{\imath}$ term coming from the addition. Now sketch the final vector. We see

$$R = \|\vec{R}\| = \sqrt{(-11.9\underline{9}5)^2 + (-3.0\underline{0}2)^2} = 12.\underline{3}6$$

$$\theta = \tan^{-1}\left(\frac{-3.0\underline{0}2}{-11.9\underline{9}5}\right) = 14.\underline{0}5°$$

WATCH OUT! Calculators output a number *in either quadrant I or IV.* By looking at my picture I can tell this is the same thing as 14.1° below the *negative* x-axis. Another way to correctly describe it is $14.1° + 180.0° = 194.1°$ from the *positive* x-axis. To avoid this confusion, I usually ignore all minus signs when determining the angle and use my sketch of the vector $\vec{R}$.

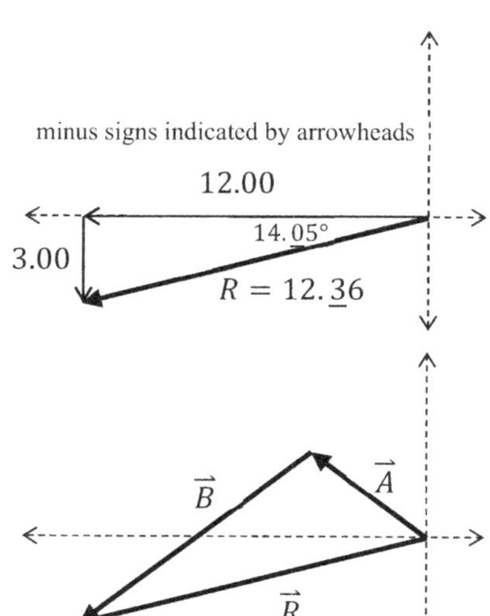

Lastly, I did the graphical vector addition just to verify it looks about right. I see my answer from above matches up fairly well as it is pointing at approximately the correct angle in the correct quadrant.

TIP: ALWAYS DO A SKETCH TO GET THE QUADRANT CORRECT.

Same example problem as previous page done with a different style

Two vectors are given as $\vec{A} = 5.00$ @ 53.1° W of N and $\vec{B} = 10.0$ @ 36.9° S of W.
Determine $\vec{R} = \vec{A} + \vec{B}$.

The angle from the positive x-axis to the vector $\vec{A}$ is $\alpha = 90.0° + 53.1° = 143.1°$.
The angle from the positive x-axis to the vector $\vec{B}$ is $\beta = 180.0° + 36.9° = 216.9°$.
Alternatively, many resources state the angle from the positive x-axis to the vector $\vec{B}$ is $\beta = -143.1°$.

Strategy if all angles are defined to positive x-axis:

1) The x-component of the vector ALWAYS USES COSINE.
2) The y-component of the vector ALWAYS USES SINE.
3) The sine & cosine functions will automatically determine the sign of the vectors for you.
4) DO NOT manually adjust signs of components after the fact based on the direction of the arrowheads.
5) DO verify the signs implied by the arrowheads match the signs in your computation.

$$\vec{A} = A_x\hat{\imath} + A_y\hat{\jmath}$$
$$\vec{A} = A\cos\alpha\,\hat{\imath} + A\sin\alpha\,\hat{\jmath}$$
$$\vec{A} = 5.00\cos 143.1°\,\hat{\imath} + 5.00\sin 143.1°\,\hat{\jmath}$$
$$\vec{A} = -3.9\underline{9}8\hat{\imath} + 3.0\underline{0}2\hat{\jmath}$$

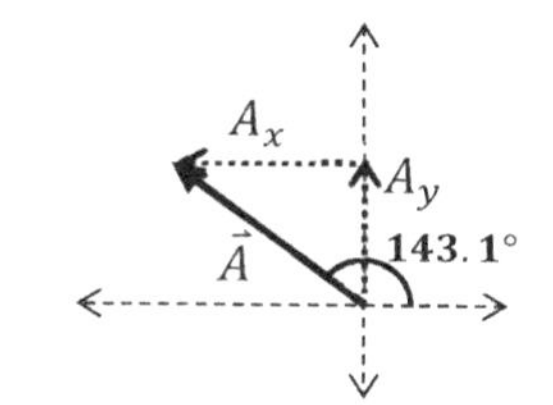

Note: correct ± signs came from computing the sine & cosine functions in your calculator.

$$\vec{B} = B_x\hat{\imath} + B_y\hat{\jmath}$$
$$\vec{B} = B\cos\beta\,\hat{\imath} + B\sin\beta\,\hat{\jmath}$$
$$\vec{B} = 10.0\cos(-143.1°)\,\hat{\imath} + 10.0\sin(-143.1°)\,\hat{\jmath}$$
$$\vec{B} = -7.9\underline{9}7\hat{\imath} + -6.0\underline{0}4\hat{\jmath}$$

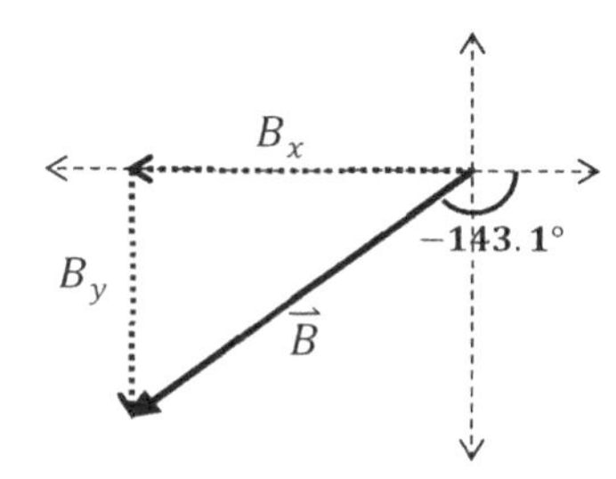

Note: correct ± signs came from computing the sine & cosine functions in your calculator.

Now we see

$$\vec{A} = -3.9\underline{9}8\hat{\imath} + 3.0\underline{0}2\hat{\jmath}$$
$$+\vec{B} = -7.9\underline{9}7\hat{\imath} - 6.0\underline{0}4\hat{\jmath}$$
$$\vec{R} = -11.9\underline{9}5\,\hat{\imath} - 3.0\underline{0}2\hat{\jmath}$$

Which method is best?

I argue the first method is best when doing problems for some problems (most in chapter 5, 6, 12, etc).
At the same time, this 2nd method is preferable if you plan to do any coding with vectors.
Both are styles useful.
Learn both styles.

3.5 Bob first walks 20.0 m at 60.0° S of E, then 30.0 m at 30.0° S of W, then an unknown direction and distance. Bob's final position is 5.00 m due east of his initial position. Use component wise addition to determine the third leg of the trip (distance and direction). Check your work by sketching the graphical addition on the grid below. Note: In this example, things won't line up perfectly on the gird lines. Let each tick mark below indicate 5 m.

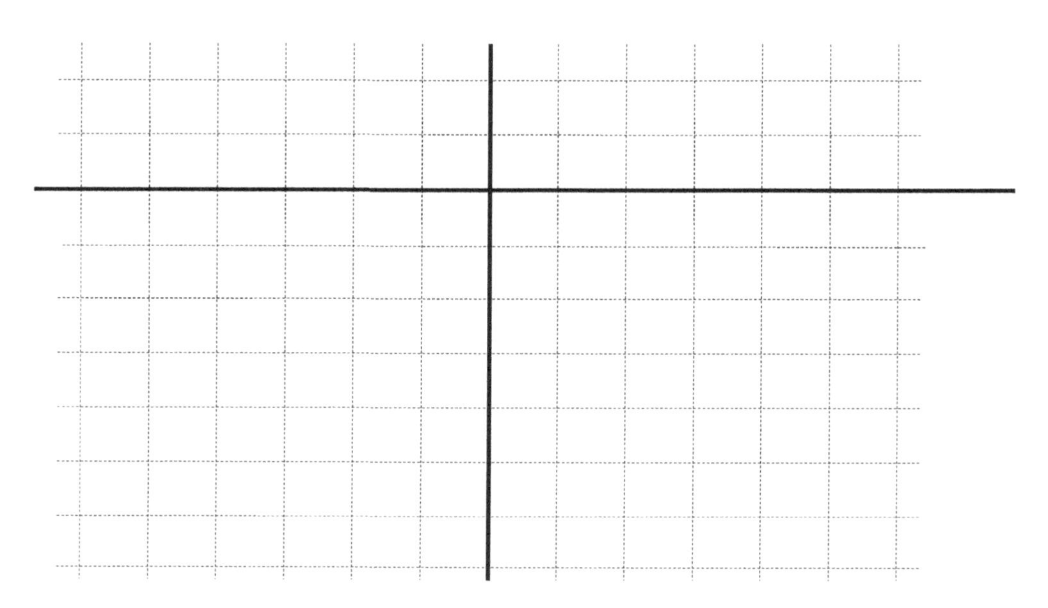

3.6 Add the following vectors: $\vec{A} = 8.00$ @ 30.0° N of W, $\vec{B} = 10.0$ @ 40.0° W of S. Express your result in both Cartesian and polar forms. Include a sketch showing the graphical vector addition.

3.7 You are told $\vec{R} = \vec{A} + \vec{B} + \vec{C}$. $\vec{A} = 8.00$ *due north*, $\vec{B} = 6.00$ @ 30.0° E of S, and $\vec{R} = 10.0$ *due west*. Determine the unknown vector $\vec{C}$. The usual Cartesian, polar, and graphical answers are expected.

3.7½ For this problem be very careful to distinguish between force *vectors* and force *magnitudes*. Two force vectors ($\vec{F}_1$ & $\vec{F}_2$) have identical magnitude F. The first force is aligned with the positive x-axis. The second can be applied at any angle θ between 0° & 180°.

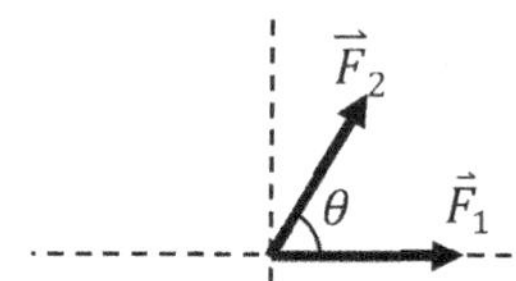

a) At what angle should $\vec{F}_2$ be placed to cause a net force magnitude of $1.75F$?
   Hint: first pretend you actually know the angle and determine $\vec{F}_{NET} = \vec{F}_1 + \vec{F}_2$ as usual. Next determine the magnitude as usual. At this point you have a relationship between the net force magnitude (given in problem statement) and the angle (what to find).
b) What direction does the net force vector point when $F_{NET} = 1.75F$?
c) At what angle should $\vec{F}_2$ be placed to cause a net force magnitude of $\frac{F}{2}$?
d) At what angle should $\vec{F}_2$ be placed to cause a net force magnitude of F?
e) What angle gives no net force?
f) What is the maximum possible net force magnitude? What angle should be used to cause maximum net force?

3.8 The Story of Mr. Boanventur…

A materials engineering student named Mr. Boanventur went out into the middle of a large flat field on a very foggy day. He proceeded to smoke a strain of medical marijuana known as 3SL. This strain was peculiar. Anyone who smoked it began to walk three straight line displacements in a manner not at all unlike an spooky movie plot. Mr. Boanventur smoked a lot of it.

At first Boanventur wasn't too messed up. For the first leg of the trip he dutifully noted he walked 100.0 m heading 36.9° N of E. Then the drug really kicked in. For the next leg of his trip he was careful to record the distance traveled as 80.0 m but totally forgot to measure his heading. On the last leg, Boanventur felt so bad about forgetting the angle that's all he could think about. He recorded the heading of his last displacement as 20.0° W of S. Of course, this time, wouldn't you know it, he forgot to record the distance traveled. Note: at some point he fell and scratched his face in a manner that made him look like Harry Potter with a beard.

Miraculously, Boanventur somehow made it back to the starting position despite the thick fog. He found a bag of crisps and began to cry tears of joy. He proceeded to apply to a prestigious local university and accidentally misspelled his name on his transfer application…for real.

Determine the heading of the second leg of the journey as well as the length of the second part of the journey. Is it impossible to determine? The vector equation gives us two equations (one eqt'n each for x- & y-directions) and we only have two unknowns…

DOT PRODUCTS

The equations and useful facts for dot products are given in the table below.

$$\vec{A}\cdot\vec{B} = \left(A_x\hat{\imath} + A_y\hat{\jmath} + A_z\hat{k}\right)\cdot\left(B_x\hat{\imath} + B_y\hat{\jmath} + B_z\hat{k}\right) = A_xB_x + A_yB_y + A_zB_z \qquad 3.1$$

$$\vec{A}\cdot\vec{B} = AB\cos\theta_{AB} \qquad 3.2$$

$$\theta_{AB} = \text{the angle between } \vec{A} \,\&\, \vec{B} \qquad 3.3$$

$$\text{The magnitude of a vector is } A = \sqrt{\vec{A}\cdot\vec{A}} = \sqrt{A_x^2 + A_y^2 + A_z^2} \qquad 3.4$$

$$\hat{\imath}\cdot\hat{\imath} = 1 \qquad \hat{\imath}\cdot\hat{\jmath} = 0 \qquad \hat{\jmath}\cdot\hat{\imath} = 0 \qquad \hat{\jmath}\cdot\hat{\jmath} = 1 \qquad \text{etc, etc} \qquad 3.5$$

The dot product of two perpendicular vectors is 0. — 3.6

The order of vectors in a dot product doesn't matter. — 3.7

The result of a dot-product is a **scalar**. — 3.8

How to Turn Any Vector into a Unit Vector

In physics, a unit vector is a convenient way to express the direction of a force in 3D space. To turn a vector into a unit vector follow these steps:

a) Determine the *magnitude* of the vector using $A = \sqrt{A_x^2 + A_y^2 + A_z^2}$.

b) Divide the original *vector* by its *magnitude*. The unit vector is $\hat{A} = \frac{\vec{A}}{A}$.

The units cancel out & the magnitude of the new *unit* vector $\hat{A}$ is 1.

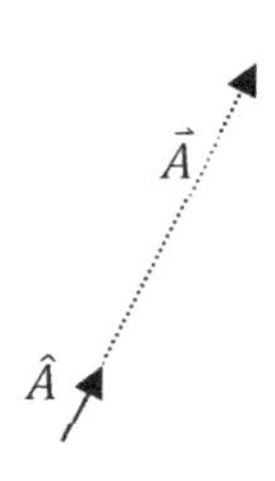

How to Determine the Angle Between Two Known Vectors

STYLE 1	STYLE 2
$\vec{A}\cdot\vec{B} = \vec{A}\cdot\vec{B}$	$\vec{A}\cdot\vec{B} = \vec{A}\cdot\vec{B}$
$AB\cos\theta_{AB} = A_xB_x + A_yB_y + A_zB_z$	$AB\cos\theta_{AB} = \vec{A}\cdot\vec{B}$
$\cos\theta_{AB} = \dfrac{A_xB_x + A_yB_y + A_zB_z}{AB}$	$\cos\theta_{AB} = \dfrac{\vec{A}\cdot\vec{B}}{AB}$ **OR** $\hat{A}\cdot\hat{B}$
$\theta_{AB} = \cos^{-1}\left(\dfrac{A_xB_x + A_yB_y + A_zB_z}{AB}\right)$	$\theta_{AB} = \cos^{-1}\left(\dfrac{\vec{A}\cdot\vec{B}}{AB}\right)$ **OR** $\cos^{-1}(\hat{A}\cdot\hat{B})$

Angle between a vector and an axis

The above procedure can be used to determine the angle between a vector and any axis.

To determine the angle between $\vec{A}$ and the *positive* x-axis let $\vec{B} = \hat{\imath}$ and do the procedure above.

$$\theta_{pos\,x\,axis} = \cos^{-1}\left(\frac{\vec{A}\cdot\hat{\imath}}{A(1)}\right) = \cos^{-1}\left(\frac{A_x}{A}\right)$$

Another way to derive the above result is to use $\hat{A}\cdot\hat{\imath} = \cos\theta_{pos\,x\,axis}$.

Notice: $\frac{A_x}{A}$ is simply the $\hat{\imath}$ part of $\hat{A}$. We could instead write the above result as

$$\theta_{pos\,x\,axis} = \cos^{-1}(\text{the } \hat{\imath} \text{ part of } \hat{A})$$

Tip: if a problem asks for the angle between a vector and the *negative* x-axis we would use

$$\hat{A}\cdot(-\hat{\imath}) = \cos\theta_{neg\,x\,axis}$$

$$\theta_{neg\,x\,axis} = \cos^{-1}(-1 \text{ times the } \hat{\imath} \text{ part of } \hat{A})$$

Example of *dot* product for two vectors in *polar* form

Two vectors are $\vec{A} = 3.00\text{ m} @ 20.0°$ west of south and $\vec{B} = 4.00\text{ m} @ 40.0°$ east of north.

I want to use the formula $\vec{A} \cdot \vec{B} = AB\cos\theta_{AB}$.

I know $A = \|\vec{A}\|$ = the magnitude of $\vec{A}$=3.00 m and $B = \|\vec{B}\|$ = the magnitude of $\vec{B}$=4.00 m.

To get the angle between I draw a quick sketch as shown below.

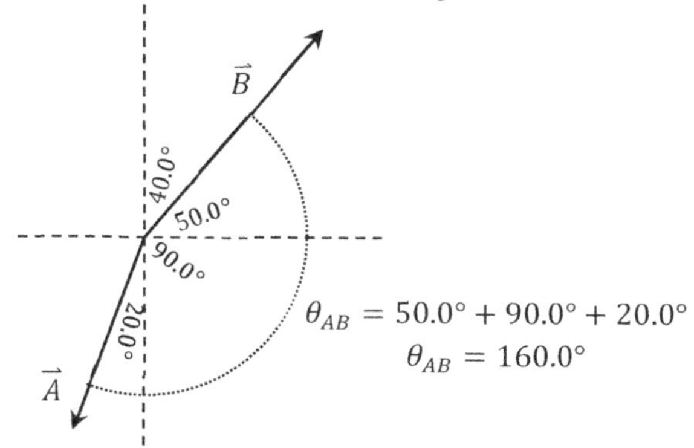

$$\theta_{AB} = 50.0° + 90.0° + 20.0°$$
$$\theta_{AB} = 160.0°$$

Now I use the formula to get

$$\vec{A} \cdot \vec{B} = AB\cos\theta_{AB}$$
$$\vec{A} \cdot \vec{B} = (3.00\text{ m})(4.00\text{ m})\cos(160.0°)$$
$$\boldsymbol{\vec{A} \cdot \vec{B} = -11.\underline{2}8\text{ m}^2}$$

Notice: we input two vectors and the output is a scalar.

Here the vectors are pointing in nearly *opposite* directions…the negative result makes sense.

Example of *dot* product for two vectors in *Cartesian* form

Two vectors are $\vec{C} = (1.00\hat{\imath} - 2.00\hat{\jmath} + 3.00\hat{k})\text{N}$ and $\vec{D} = (0.00\hat{\imath} - 4.00\hat{\jmath} - 2.00\hat{k})\text{m}$.

$$\vec{C} \cdot \vec{D} = C_x D_x + C_y D_y + C_z D_z$$
$$\vec{C} \cdot \vec{D} = \{(1.00)(0.00) + (-2.00)(-4.00) + (3.00)(-2.00)\}\text{N} \cdot \text{m}$$
$$\boldsymbol{\vec{C} \cdot \vec{D} = 2.00\text{ N} \cdot \text{m}}$$

Notice: with 3D vectors it becomes much more important to trust the math.

3.9 Dot Product Practice

You are given four vectors in the table at right.

a) Determine $\vec{A} \cdot \vec{B}$.

b) Determine $\vec{C} \cdot \vec{D}$.

c) Determine the *unit vector* $\hat{A}$ in Cartesian form.

d) Determine the *unit vector* $\hat{C}$ in Cartesian form.

e) Determine the angle between $\vec{A}$ & $\vec{B}$.

f) Determine the angle between $\vec{C}$ & $\vec{D}$.

g) Determine the angle between $\vec{C}$ and the *positive* z-axis.

h) Determine $\vec{A} \cdot \vec{C}$.

i) Determine the angle between $\vec{B}$ & $\vec{D}$.

j) Determine the angle between $\vec{D}$ and the *negative* y-axis.

k) Determine the angle between $\vec{B}$ and the *positive* z-axis.

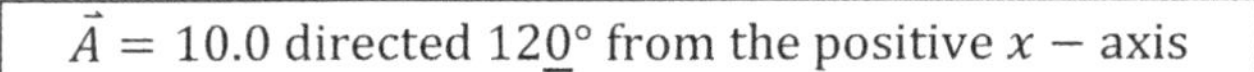
$\vec{A} = 10.0$ directed $12\underline{0}°$ from the positive $x -$ axis

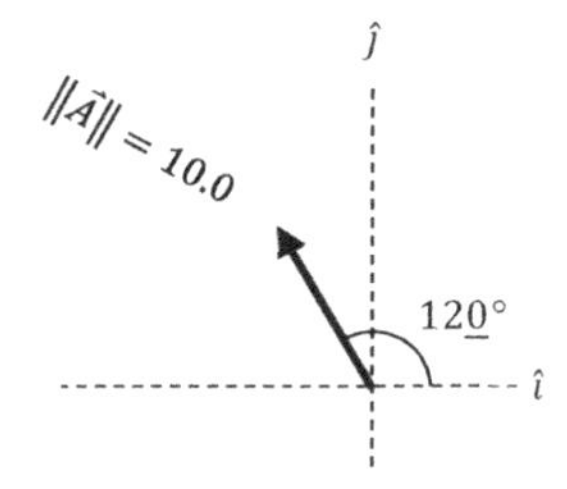

$\vec{B} = 20.0$ directed $-30.0°$ from the positive $x -$ axis

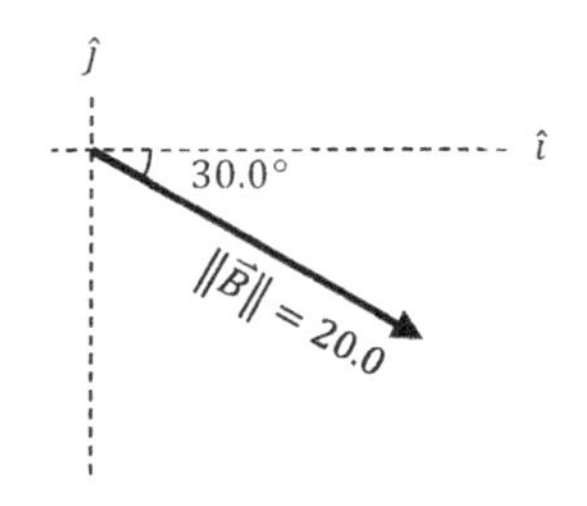

$\vec{C} = 1.00\hat{\imath} - 2.00\hat{\jmath} + 3.00\hat{k}$

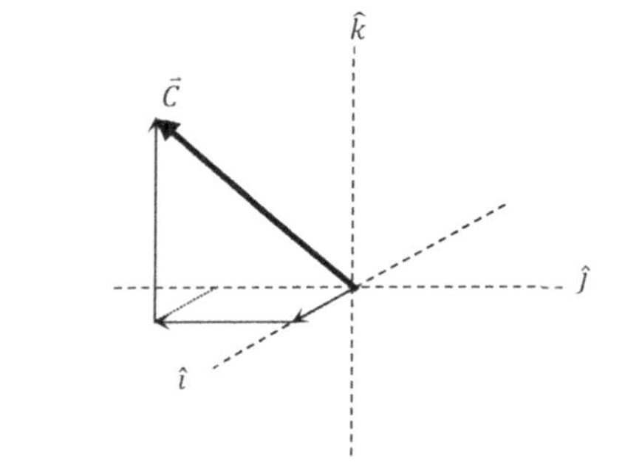

$\vec{D} = -3.00\hat{\imath} + 1.00\hat{\jmath} - 2.00\hat{k}$

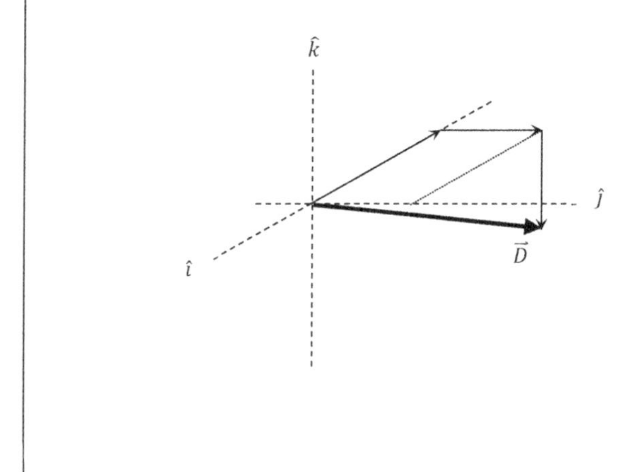

CROSS PRODUCTS

The equations and useful facts for cross-products are given below

$$\vec{A} \times \vec{B} = (A_x\hat{\imath} + A_y\hat{\jmath} + A_z\hat{k}) \times (B_x\hat{\imath} + B_y\hat{\jmath} + B_z\hat{k}) \quad 3.9$$

The magnitude of the cross-product is $\|\vec{A} \times \vec{B}\| = AB \sin\theta_{AB}$
Note: since θ_{AB} is always between $0°$ & $180°$ we know $\sin\theta_{AB} > 0$. 3.10

Use the Right Hand Rule to Determine Direction of Cross-Product
1) Align fingers of right hand with first vector. 3.11
2) Curl fingers of right hand towards second vector.
3) Thumb points in direction of result.

$\hat{\imath} \times \hat{\imath} = 0$ $\quad$ $\hat{\imath} \times \hat{\jmath} = \hat{k}$ $\quad$ $\hat{\jmath} \times \hat{\imath} = -\hat{k}$ $\quad$ $\hat{\jmath} \times \hat{\jmath} = 0$ $\quad$ etc, etc 3.12

The cross product of two parallel (or anti-parallel) vectors is 0. 3.13

Switching the order of vectors in cross-product flips the sign of the result. 3.14

The magnitude of $\vec{A} \times \vec{B}$ is the area of the parallelogram defined by $\vec{A}$ & $\vec{B}$. 3.15

The result of a cross-product is a **vector**. 3.16

In 3.12, it is difficult to keep track of signs. I remember the signs using a trick I call the wheel of pain (shown at right). You use the wheel of pain to figure out the sign of the cross product between two unit vectors.

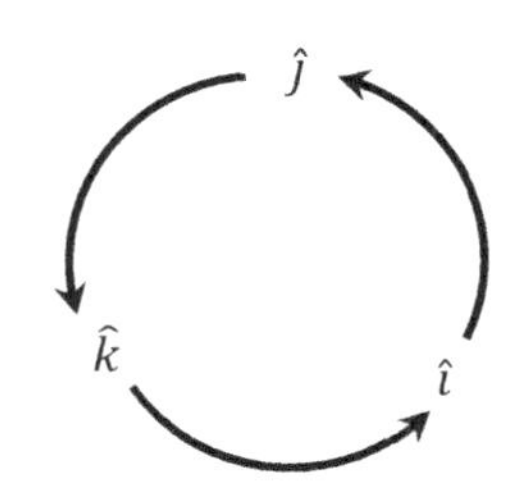

For example, consider $\hat{\imath} \times \hat{k}$. The first term in the cross product is $\hat{\imath}$ so start there in the wheel of pain. The next term is $\hat{k}$ so move around the wheel in that direction. The next term around the wheel is the result (with a ± sign). Here the result of the cross product is $\hat{\imath} \times \hat{k} = -\hat{\jmath}$. The minus sign comes from the fact that we were going opposite the arrows of the wheel of pain.

To practice using the wheel, verify that $\hat{\jmath} \times \hat{k} = \hat{\imath}$ while $\hat{k} \times \hat{\jmath} = -\hat{\imath}$.

Using the wheel of pain to determine cross products is pretty darn fast if one or more of the components in either vector is zero. This is common in physics problems as we are usually free to align our coordinates with one of the vectors and set two of the three vector components to zero.

Example of *cross* product for two vectors in *polar* form

Two vectors are $\vec{A} = 3.00\text{ m @ }20.0°$ west of south and $\vec{B} = 4.00\text{ m @ }40.0°$ east of north.

I want to use the formula $\|\vec{A} \times \vec{B}\| = AB\sin\theta_{AB}$ to get the magnitude of the cross product.

I must also use the right hand rule to get the direction of the cross product.

Remember: the output of the cross product is a vector so we need both magnitude and direction.

I know $A = 3.00$ m and $B = 4.00$ m.

To get the angle between I draw a quick sketch as shown below.

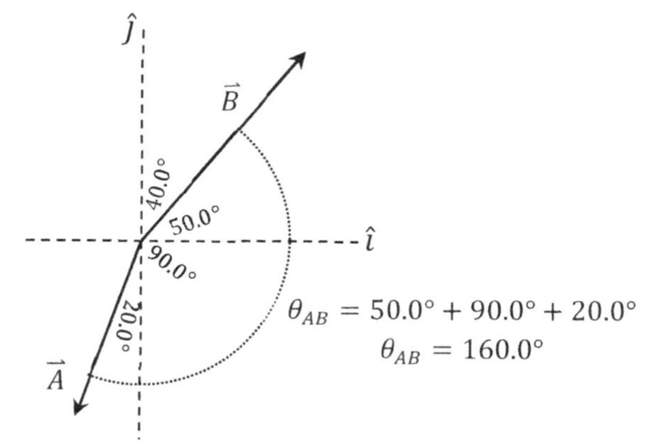

$$\theta_{AB} = 50.0° + 90.0° + 20.0°$$
$$\theta_{AB} = 160.0°$$

Now I use the formula to get

$$\|\vec{A} \times \vec{B}\| = AB\sin\theta_{AB}$$
$$\|\vec{A} \times \vec{B}\| = (3.00\text{ m})(4.00\text{ m})\sin(160.0°)$$
$$\|\vec{A} \times \vec{B}\| = 4.10\text{ m}^2$$

Here the vectors are pointing in nearly *opposite* directions…the result close to zero makes sense.

Remember: the *cross* product is zero when the input vectors are parallel (or anti-parallel).

Finally, I use the right hand rule to determine the direction of the cross product.

1) I line up the fingers of my right hand with the first vector in the cross-product (in this case, $\vec{A}$).
2) I curl the fingers of my right hand to the second vector (in this case, curl to $\vec{B}$).
3) I observe the thumb of my right hand points out of the page ($+\hat{k}$ according to this coordinate system).

The final *vector* result of the cross product is thus

$$\boldsymbol{\vec{A} \times \vec{B} = 4.10\text{ m}^2(+\hat{k})}$$

Example of *cross* product for two vectors in *Cartesian* form using the wheel of pain

Two vectors are $\vec{C} = (1.00\hat{\imath} - 2.00\hat{\jmath} + 3.00\hat{k})\text{N}$ and $\vec{D} = (0.00\hat{\imath} - 4.00\hat{\jmath} - 2.00\hat{k})\text{m}$.

$$\vec{C} \times \vec{D} = (1.00\hat{\imath} - 2.00\hat{\jmath} + 3.00\hat{k})\text{N} \times (0.00\hat{\imath} - 4.00\hat{\jmath} - 2.00\hat{k})\text{m}$$

Note: I know terms with $\hat{\jmath} \times \hat{\jmath} = 0$ and $\hat{k} \times \hat{k} = 0$ will drop out…so I don't even bother to write them.

Also, I take great care to not mess up the order of multiplication…it matters for a *cross* product!

$$\vec{C} \times \vec{D} = \{(1.00\hat{\imath}) \times (-4.00\hat{\jmath}) + (1.00\hat{\imath}) \times (-2.00\hat{k}) + (-2.00\hat{\jmath}) \times (-2.00\hat{k}) + (3.00\hat{k}) \times (-4.00\hat{\jmath})\}\text{N·m}$$
$$\vec{C} \times \vec{D} = \{-4.00(\hat{\imath} \times \hat{\jmath}) - 2.00(\hat{\imath} \times \hat{k}) + 4.00(\hat{\jmath} \times \hat{k}) - 12.0(k \times \hat{\jmath})\}\text{N·m}$$

I use the wheel of pain to determine $\hat{\imath} \times \hat{\jmath} = +\hat{k}$, $\hat{\imath} \times \hat{k} = -\hat{\jmath}$, $\hat{\jmath} \times \hat{k} = +\hat{\imath}$, and finally $\hat{k} \times \hat{\jmath} = -\hat{\imath}$.

$$\vec{C} \times \vec{D} = \{-4.00(+\hat{k}) - 2.00(-\hat{\jmath}) + 4.00(+\hat{\imath}) - 12.0(-\hat{\imath})\}\text{N·m}$$

Now I group like terms and reorder the sequence in stand format ($\hat{\imath}$ first, then $\hat{\jmath}$, then $\hat{k}$).

$$\boldsymbol{\vec{C} \times \vec{D} = (-8.00\hat{\imath} + 2.00\hat{\jmath} - 4.00\hat{k})\text{N·m}}$$

3.10 Before doing any cross products, you must first learn to recognize if a coordinate system is *right-handed.* A right handed coordinate system is one which conforms to the wheel of pain. To check this, line up the fingers of your right hand with $\hat{\imath}$, curl your fingers to $\hat{\jmath}$, and verify your thumb actually points in the positive $\hat{k}$ direction. If your thumb aligns with the positive $\hat{k}$ direction, the coordinate system is right-handed. Determine which of the following coordinate systems are right handed.

In Systems E through H the ⊙ symbol implies "out of the page" while the ⊗ symbol implies into the page. I remember it this way: the dot is an arrowhead coming out of the page while the × is the feathery tail of the arrow flying into the page.

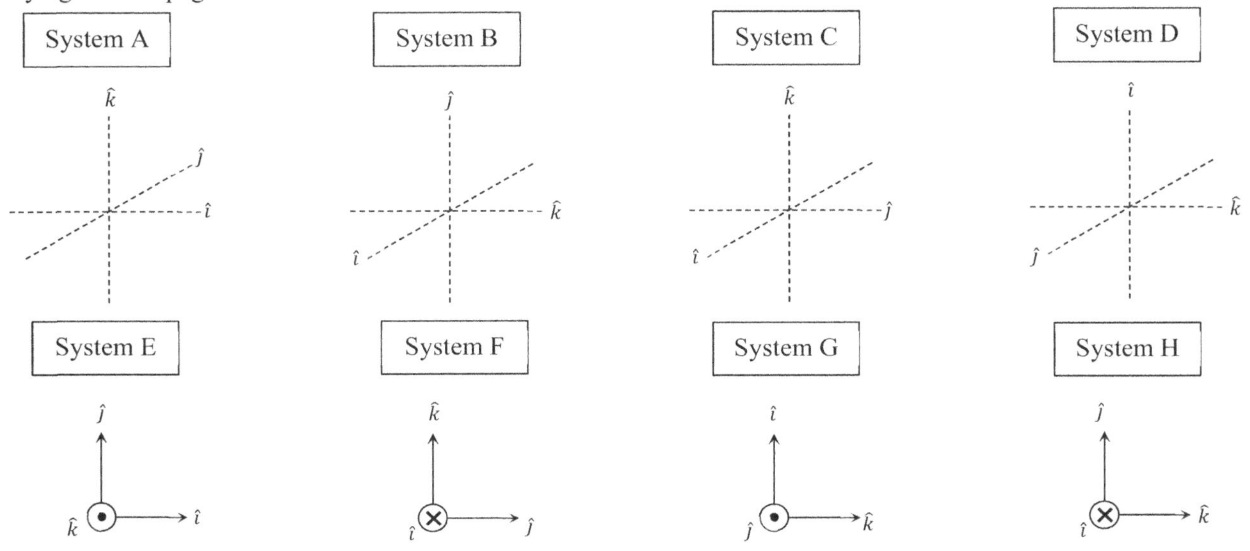

3.11 Cross Product Practice

You are given three vectors in the table below. Note: you should immediately verify the coordinate system is right-handed before proceeding. Figures are *approximately* to scale but it isn't perfect.

a) Determine $\vec{A} \times \vec{B}$.

b) Determine $\vec{C} \times \vec{B}$.

c) Why it is possible to compute $\vec{C} \cdot (\vec{A} \times \vec{B})$ but not $\vec{C} \times (\vec{A} \cdot \vec{B})$?

d) Determine $\vec{B} \times \vec{A}$.

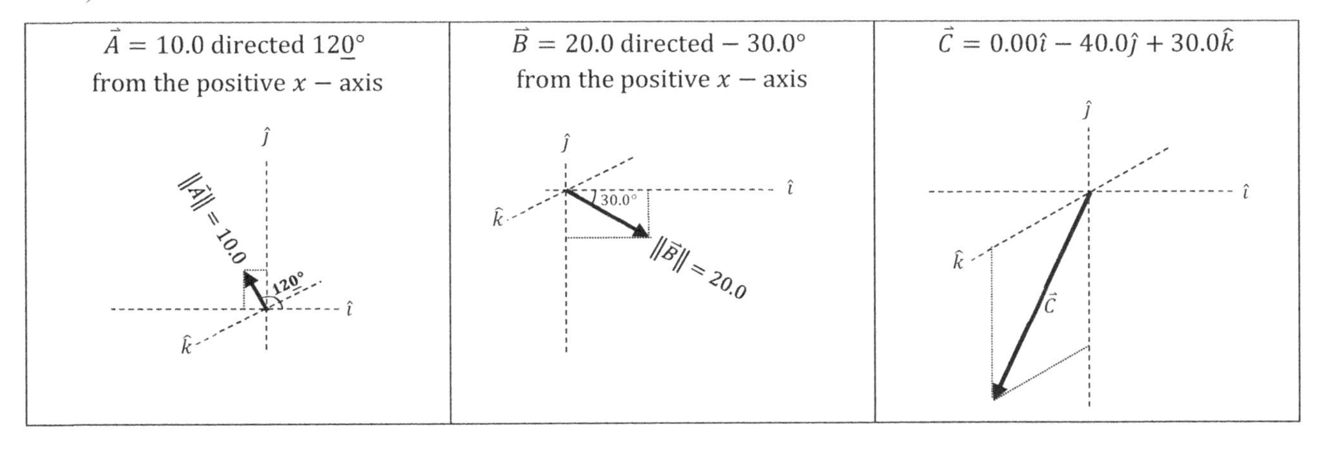

$\vec{A} = 10.0$ directed $120°$ from the positive $x-$axis	$\vec{B} = 20.0$ directed $-30.0°$ from the positive $x-$axis	$\vec{C} = 0.00\hat{\imath} - 40.0\hat{\jmath} + 30.0\hat{k}$

3.12 A drone, initially at the origin is repositioned using three displacements. The first displacement, shown in black, moves the drone 10.00 m heading 36.87° south of east. The second displacement, shown in grey, moves the drone upwards 7.25 m. The final position of the drone, shown in white, is in the plane defined by the east-west axis and the vertical axis with the angle shown in the figure. The final distance from the origin is 5.00 m.

a) Write down the 1st displacement in Cartesian form with *four* sig figs.
b) Write down the final position vector in Cartesian form with *four* sig figs.
c) What 3rd displacement vector was required to give the drone the stated final position? Answer in Cartesian form with *three* sig figs.

3.13 A position vector is given as $\vec{r} = -3.00\hat{\imath} + 2.00\hat{\jmath}$. You may assume the units on this position vector are meters. A *linear* momentum vector is given by $\vec{p} = 5.00\hat{\imath} - 4.00\hat{k}$. The assumed units on the momentum vector are $\text{kg}\cdot\frac{\text{m}}{\text{s}}$. *Angular* momentum ($\vec{L}$) is defined by the equation

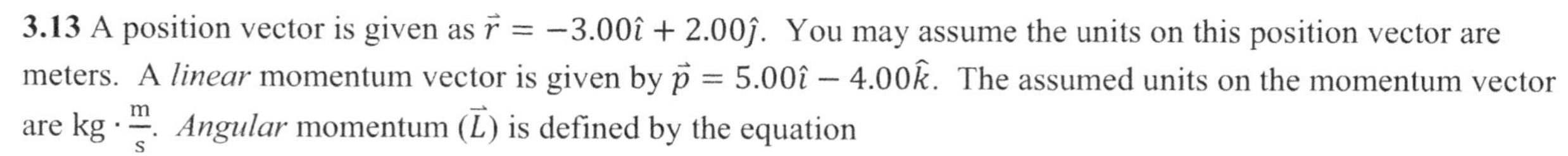

$$\vec{L} = \vec{r} \times \vec{p}$$

a) Determine the units appropriate for *angular* momentum.
b) Determine the angle between the position vector and the *linear* momentum vector.
c) Determine the *angular* momentum vector by computing the cross-product. Answer in Cartesian form and keep three sig figs on all components. Don't forget to include the units you determined in part a!!!

3.14 An electric field vector is $\vec{E} = -2.00\hat{\imath} + 3.00\hat{\jmath} - 4.00\hat{k}$. For this problem I will ignore the units. If you care, the units are either $\frac{\text{V}}{\text{m}}$ (volts per meter) or $\frac{\text{N}}{\text{C}}$ (newtons per coulomb).

a) Determine the magnitude of the electric field vector. Tip: when the problem statement is written with decimals, that implies you should answer with as a decimal with the same number of sig figs.
b) Determine a unit vector describing the direction of the electric field. Each term should have a decimal number with three sig figs.
c) Determine the angle between the electric field and the positive z-axis. Again, assume I want three sig figs for final answer.

3.15 The displacement vector between random points **A** and **B** is given by

$$\vec{r} = 2.00\hat{\imath} - 3.00\hat{\jmath} - 4.00\hat{k}$$

A force vector is

$$\vec{F} = -5.00\hat{\imath} + 6.00\hat{k}$$

Note: the units on $\vec{r}$ are meters (m) and the units on $\vec{F}$ are Newtons (N).

a) Write down a unit vector pointing in the same direction as $\vec{r}$. Answer in Cartesian form with 3 sig figs for each term.
b) Determine the angle between $\vec{r}$ and the positive z-axis. Express your result as a number between 0° and 180°.
c) Torque is defined as $\vec{\tau} = \vec{r} \times \vec{F}$. Determine the torque using the given vectors. Answer in Cartesian form.

3.16 A moth starts from rest and undergoes three displacements. The first displacement is $\vec{A}$ shown in white while the second is $\vec{B}$ shown in black. Vector $\vec{A}$ has magnitude 10 and lies in the yz-plane angled 20° from the z-axis. Vector $\vec{B}$ has magnitude 20 and lies in the xy-plane angled 30° from the y-axis. After a *third* displacement, the moth has returned to the origin.

a) Determine the 3rd displacement in *Cartesian* form.
b) Determine the *magnitude* of the third displacement.
c) Determine the *angle* the third displacement makes with the negative z-axis.

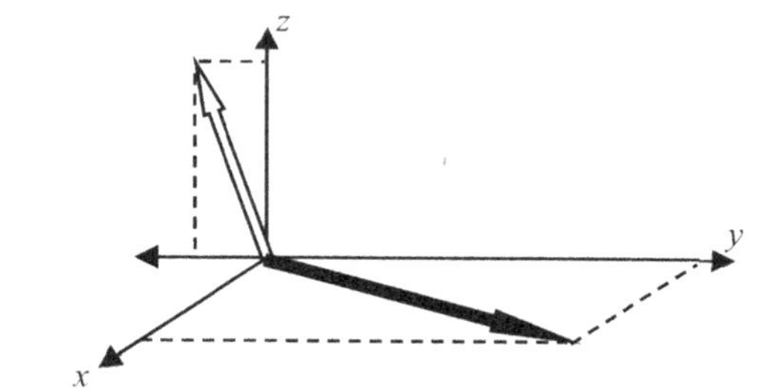

3.17 A power line structure is secured by three guy lines. The three guys are symmetrically located about the base of the structure. Assume the point where the structure touches the ground is the origin. For simplicity, assume all guy lines attach to the structure at the point **D** 20.0 m above the origin. The point **A** lies on the positive x-axis while the axis of the structure lies on the positive z-axis. The positive y-axis is to the right in the figure. The guys are each 35.0 m long.

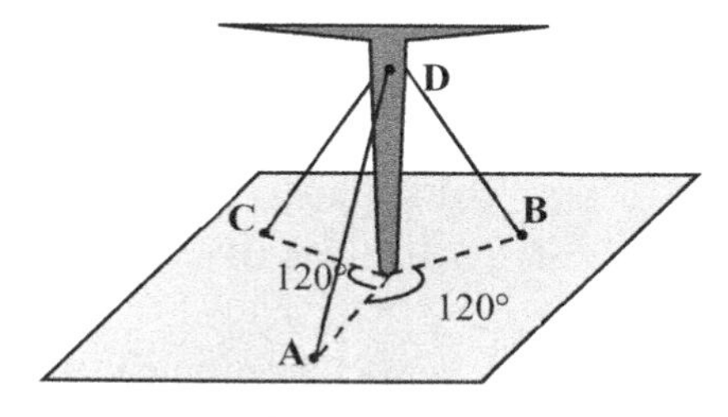

a) Determine the unit vector $\hat{A}$ pointing from the attachment point to point **A**. Verify $\hat{A}$ is indeed a unit vector by determining its magnitude and showing $\|\hat{A}\| = \sqrt{\hat{A}\cdot\hat{A}} = 1$.
b) Determine the unit vector pointing from the attachment point to point **B**.
c) Determine the unit vector pointing from the attachment point to point **C**.
d) Determine the angle between any two guy lines.
e) **Challenge:** The wind picks up and causes a force of 1000 N that acts at point **D** pointing parallel to the positive y-axis. The tension in guy **C** is 6000 N. The ground pushes straight up (parallel to the z-axis) with unknown magnitude Z. The vector sum of all forces acting on the structure is zero. Determine the tension in guys **A** and **B** and the magnitude of the upwards force from the ground.

This type of computation is could also be used to determine bonding angles in chemistry or in studying crystal lattices in a materials class.

3.18 In a grocery store a bunch of oranges are stacked for display. The bottom plane of oranges in the stack forms a square. Assume each orange has identical radius R. Notice that a coordinate system is shown. In particular note that for *this picture* right and left relate to $\hat{\jmath}$. The center of each of the bottom oranges are in the xy-plane. Some oranges are numbered to ease communication.

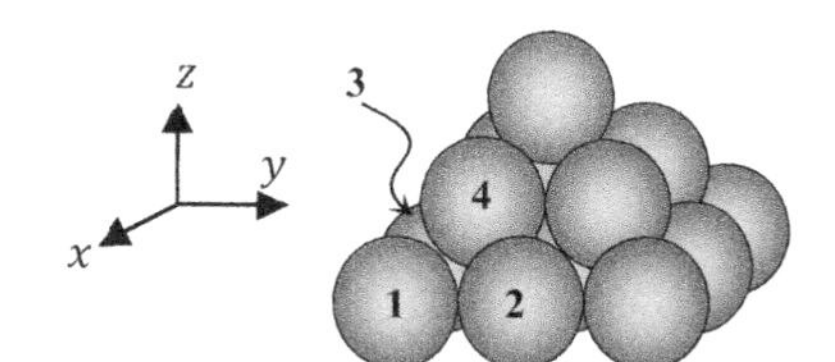

a) Determine a vector pointing from of orange 1 to orange 2 (center-to-center).
b) Determine a vector pointing from orange 1 to orange 3 (center-to-center).
c) Determine a vector pointing from orange 1 to orange 4 (center-to-center).

Note: we could change the bottom plane of oranges in the stack to form an equilateral triangle or hexagon. In these orientations we say the oranges are in Face Centered Cubic (FCC) or hexagonal close pack (HCP) respectively. It turns out atoms will pack most efficiently (least amount of empty space) in HCP. Also interesting, FCC and HCP are actually the same lattice structure if one is clever about how you rotate the coordinate system! Lattice structure relates to many properties of materials, for instance electrical conductivity. More on this in Condensed Matter physics courses (previously known as Solid State Physics) or Materials courses.

3.19 The quantum mechanical spin of nuclei causes them to act like small magnets. The magnetism is characterized by the magnetic moment ($\vec{\mu}$) which indicates the size and orientation of the magnetism. When exposed to an external magnetic field $(\vec{B})$ the potential energy (U) associated with the molecule in the field is given by

$$U = -\vec{\mu} \cdot \vec{B}$$

The molecule will also experience a torque ($\vec{\tau}$) given by

$$\vec{\tau} = \vec{\mu} \times \vec{B}$$

An external magnetic field $\vec{B}$ aligned with the positive y-axis has magnitude 1.0 T. A proton at the origin has a magnetic moment with magnitude $\mu = 5.0 \times 10^{-27} \frac{\text{J}}{\text{T}}$.

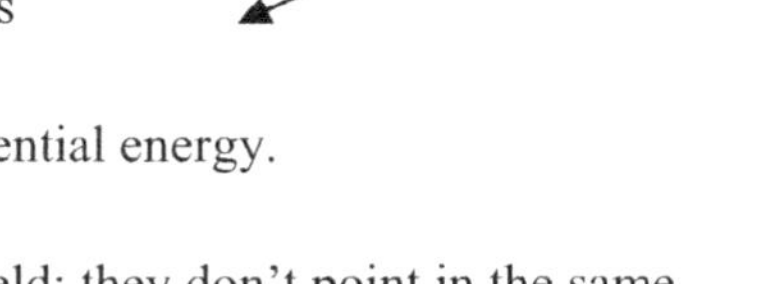

a) For what direction of $\vec{\mu}$ will the proton have <u>max</u> potential energy?
b) For what direction of $\vec{\mu}$ will the proton have <u>min</u> potential energy?
c) Determine the energy difference between the min and the max.
d) For what directions of $\vec{\mu}$ will the proton experience no torque?
e) Determine the potential energy and torque when the magnetic moment is aligned with the positive z-axis.
f) Determine all possible orientations of the proton for which it has no potential energy.

Note: In an MRI the magnetic moments align by *precessing* about the external field; they don't point in the same direction but act instead somewhat like a spinning top. While the above example isn't an accurate representation of this, it gets the idea across. Also, expect the numbers in real life to differ from my estimates.

The energy difference between the anti-aligned and aligned states corresponds to radio frequencies. By sending in a pulse of radio waves with a frequency corresponding to the energy from part c) the magnetic moments will flip from aligned to anti-aligned. When the pulse is turned off, the magnetic moments will return to the aligned state. In the process of returning to the lower energy state, the nuclei emit radio waves which can be detected and used to non-destructively determine material composition. There is quite a lot more to this but I thought this would be fun to consider.

Example: Torque is defined as $\vec{\tau} = \vec{r} \times \vec{F}$. The vector $\vec{r}$ is the displacement from the pivot point (axis of rotation) to the point where force is applied.

Consider the figure shown at right. The force of magnitude F lies in a plane parallel to the yz-plane and is applied at the end of the bent, black rod. Assuming the origin is the pivot point, determine the torque exerted by the force on the rod.

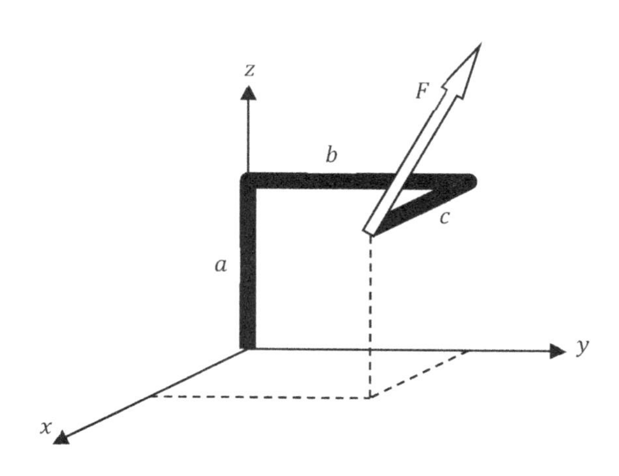

$$\vec{r} = c\hat{\imath} + b\hat{\jmath} + a\hat{k}$$
$$\vec{F} = 0\hat{\imath} + F_y\hat{\jmath} + F_z\hat{k}$$
$$\vec{\tau} = \left(c\hat{\imath} + b\hat{\jmath} + a\hat{k}\right) \times \left(F_y\hat{\jmath} + F_z\hat{k}\right)$$

Ignore terms with $\hat{k} \times \hat{k} = 0$ and $\hat{\jmath} \times \hat{\jmath} = 0$.

$$\vec{\tau} = cF_y(\hat{\imath} \times \hat{\jmath}) + cF_z\left(\hat{\imath} \times \hat{k}\right) + bF_z\left(\hat{\jmath} \times \hat{k}\right) + aF_y\left(\hat{k} \times \hat{\jmath}\right)$$

Use the wheel of pain to show

$$\vec{\tau} = cF_y\left(\hat{k}\right) + cF_z(-\hat{\jmath}) + bF_z(\hat{\imath}) + aF_y(-\hat{\imath})$$
$$\vec{\tau} = \left(bF_z - aF_y\right)\hat{\imath} - cF_z\hat{\jmath} + cF_y\hat{k}$$

Note: you could figure out F_z and F_y using SOH CAH TOA if an angle is known. I was more interested in showing you how to do the cross-product stuff.

3.20 A simple model of a utility pole is shown at right. The above-ground portion of the pole is approximately 6.00 m tall with a 2.00 m long cross-bar centered on top. The origin is located at the base of the pole which is aligned with the positive z-axis. A wire causes a force of tension at point **A** in the direction indicated with a magnitude of 2750 N. To be clear, the force of tension is in the yz-plane.

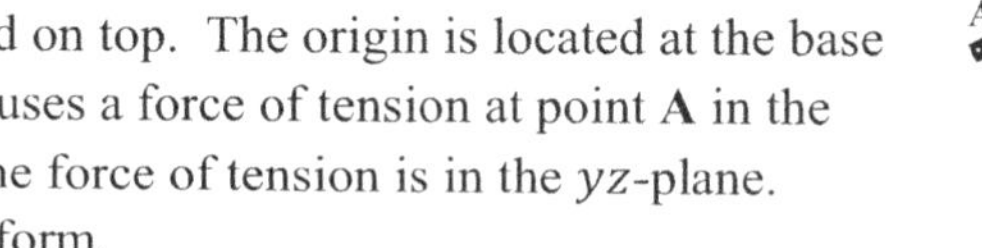

a) Determine the position vector of point **A** in Cartesian form.

b) Write the force vector $\vec{F}$ in Cartesian form.

c) The torque caused by the tension is given by $\vec{\tau} = \vec{r} \times \vec{F}$. Determine $\vec{\tau}$.

3.21 Two pole-mount transformers are attached to a utility pole. Doing a quick web search for specifications, I found a transformer with the specs shown in the figure. The cross-rod (dark grey rod) is 24 cm by 24 cm by 1.5 m. The pole is 9.0 m tall with 20 cm diameter. The top of the cross-rod attaches 1.0 m below the top of the pole. Various views shown in an attempt to make the dimensions of the problem easier to see. **Note: figures not to scale.** The center of mass of the transformer is shown in the side view labeled as **CM**.

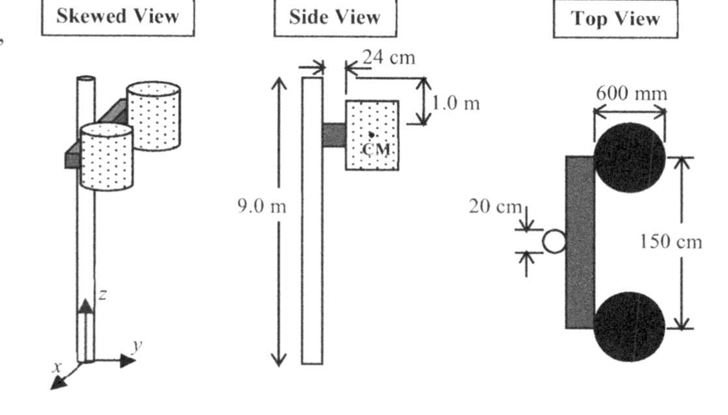

a) Assume the origin is located at the center of the pole's base. Determine the position vectors ($\vec{r}$) for the center of mass (CM) of each transformer.

b) The transformer's weight is a force with magnitude 3000 N that acts at the center of mass pointing straight down ($-\hat{k}$). The torque exerted by the transformer is given by $\vec{\tau} = \vec{r} \times \vec{F}$ where $\vec{F}$is the weight force. Determine $\vec{\tau}$ for each transformer.

Comments: I tried to use the units I found upon doing a web search. This shows how mixed units might appear in a real world design problem. Notice even a simple problem like this requires attention to detail.

3.22 Two vectors lie in the xy-plane. The angle between the two vectors is 153.44° while the dot product of the two vectors is -1000. One of the vectors has a scalar x-component of 20 while the other has a scalar x-component of -30. Determine the magnitude of each vector.

3.23 A student gives you a puzzle he claims has two solutions. The student identifies two vectors that lie in the xy-plane. He says the dot product of the two vectors is -6.000 while the cross product is 8.000. The first vector has magnitude $A = 5.000$ while the second has a scalar x-component of 1.500. He challenges you to determine two possible solutions for $\vec{A}$ and $\vec{B}$ and the angle between them.

3.24 Two balls hang from the ceiling using light strings. Each ball has static charge on it. As a result, the two balls repel from one another and hang at the angle shown in the picture. The distance of separation is d and the length of each string is L. Assume the size of each ball is negligible compared to the length of the string. Assume the origin of the coordinate system is the point where the two strings attach to the ceiling.

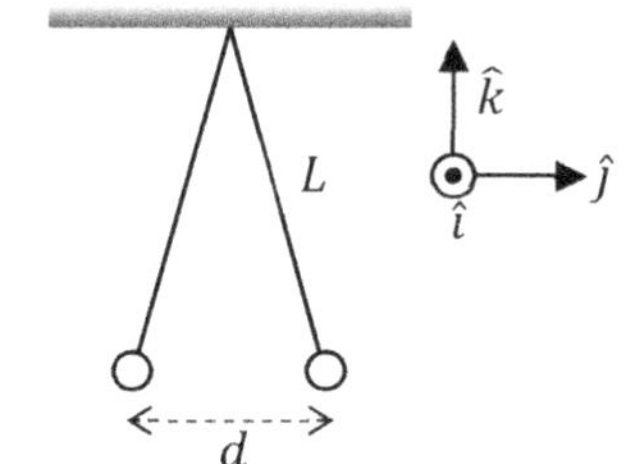

a) Determine a vector describing the displacement from the *left* ball to the origin.
b) Determine a vector describing the displacement from the *right* ball to the origin.

3.25 This time *three* balls are hanging from strings. Each ball again has static charge. The balls separate from each other and form an equilateral triangle in the horizontal plane. The side of the equilateral triangle is s and the length of each string is L. Assume the size of each ball is negligible compared to the length of the string. Assume the origin of the coordinate system is the point where the three strings attach to the ceiling.

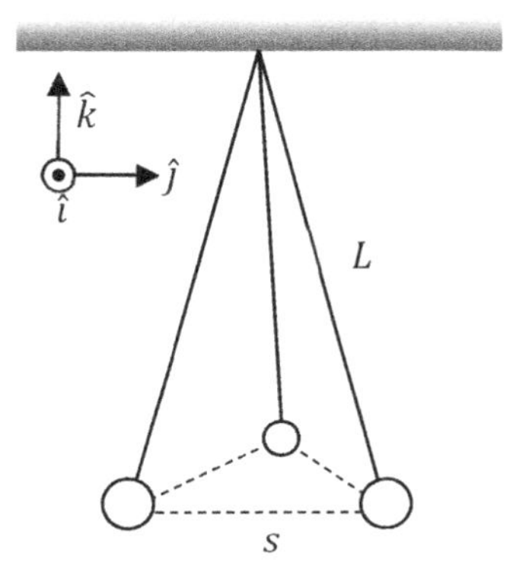

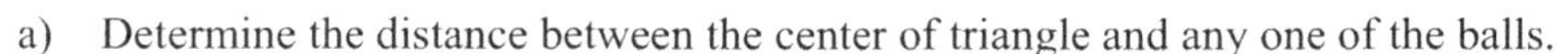

a) Determine the distance between the center of triangle and any one of the balls.
b) Determine a vector describing the displacement from the *front left* ball to the origin.
c) At what angle, from the vertical axis, does the ball hang?

PROJECTILES & 2D/3D MOTION

Suppose we have a particle moving in two dimensions instead of just one. We can write the position vector $\vec{r}(t)$ as

$$\vec{r}(t) = x(t)\hat{\imath} + y(t)\hat{\jmath}$$

The motion in each dimension is independent. This means we can solve problems just like two-stage problems from Chapter 2: one stage for the x-direction and one stage for the y-direction. For a constant acceleration problem, you could use the equations below twice (once for the x-direction and once for the y-direction).

$x_f = x_i + v_{ix}t + \frac{1}{2}a_x t^2$	$v_{fx}^2 = v_{ix}^2 + 2a_x \Delta x$	$\Delta x = \frac{1}{2}(v_{ix} + v_{fx})t$	$v_{fx} = v_{ix} + a_x t$

Independence of xy-motion

4.1 A car rolls along a flat track with no friction at constant speed. At a certain point the car launches a ball vertically. When the ball comes back down, will the ball land in front of, behind, or back on top of the car?

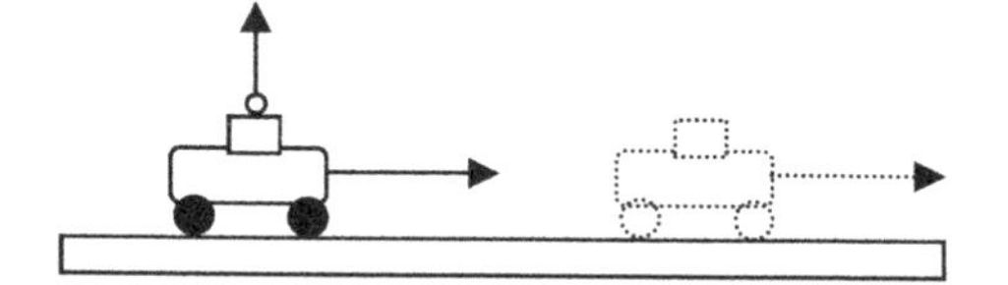

4.2 A car rolls along a flat track with no friction. It starts from rest but is pulled by a hanging mass. At a certain point the car launches a ball vertically. When the ball comes back down, will the ball land in front of, behind, or back on top of the car?

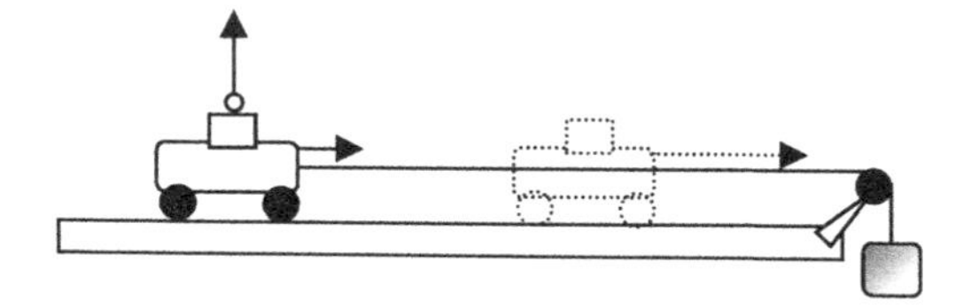

4.3 A car starts from rest and rolls down an incline. At some point the car launches a ball perpendicular to the plane. When the ball comes back down, will the ball land in front of, behind, or back on top of the car?

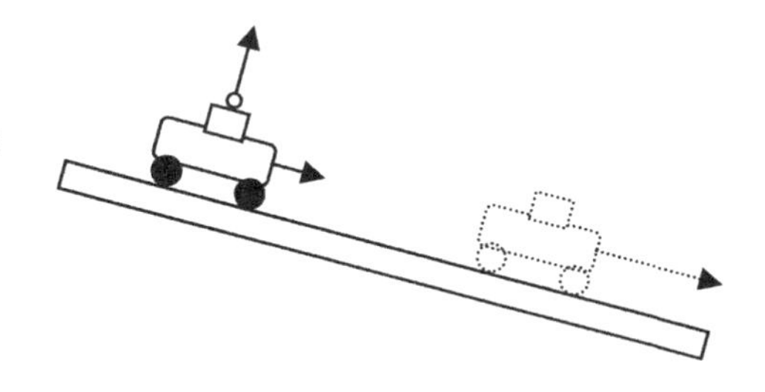

4.4-4.7 Suggested simulation activities are available on robjorstad.com in the supplemental handout. **Might need updating as sims have changed.**

<u>Standard</u> Projectile Problems:

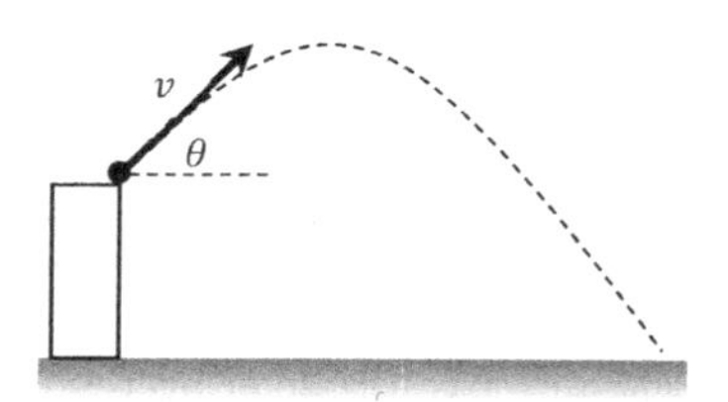

A standard projectile problem considers someone throwing/kicking a ball, a cannon shooting a projectile, etc. Standard projectiles need not have equal launch and impact positions. **Unless specifically stated, we assume air resistance is negligible.**

For all standard projectile problems the following info applies.
I am assuming UP IS THE POSITIVE DIRECTION.
NOTE: You may get some sign changes if you assume down is the positive direction.

$x_i = 0 \Rightarrow x_f = \Delta x = x$	$a_y = -g$	$y_f = y_i \pm v_i \sin\theta\, t - \frac{g}{2}t^2$
$a_x = 0$	$v_{iy} = \pm v_i \sin\theta$ Use – for downward initial launch	$v_{fy}^2 = v_{iy}^2 - 2g\Delta y$
$x = v_i \cos\theta\, t$	$\Delta y = \pm h$ Use – when final lower than initial	$v_{fy} = v_{iy} - gt$
$v_{fx} = v_{ix} = v_i \cos\theta$	**Final speed**: $v_f = \sqrt{v_{fx}^2 + v_{fy}^2}$ **Impact angle**: $\phi = \tan^{-1}\left(\frac{v_{fy}}{v_{fx}}\right)$	$y(x) = y_i + x\tan\theta - \frac{gx^2}{2v_i^2\cos^2\theta}$
At max height the velocity is $\vec{v} = v_{ix}\hat{\imath} + 0\hat{\jmath}$. Only for a purely vertical launch is speed zero at max height.		

I want to point out an important distinction between

$$y_f = y_i \pm v_i \sin\theta\, t - \frac{g}{2}t^2 \qquad \& \qquad y(x) = y_i + x\tan\theta - \frac{gx^2}{2v_i^2\cos^2\theta}$$

- The first gives vertical position versus *time*; the second gives vertical position versus *horizontal position.*
- The first curve is a parabola but it is <u>not</u> the shape of the object's flight path.
- The second curve <u>is</u> the actual shape of the projectile's flight path…the trajectory through space.
- The derivative of the first equation gives the y-component of velocity as a function of time.
- The derivative of the second relates to the *angle* of the velocity as a function of horizontal position.

Standard Projectile with Level Ground

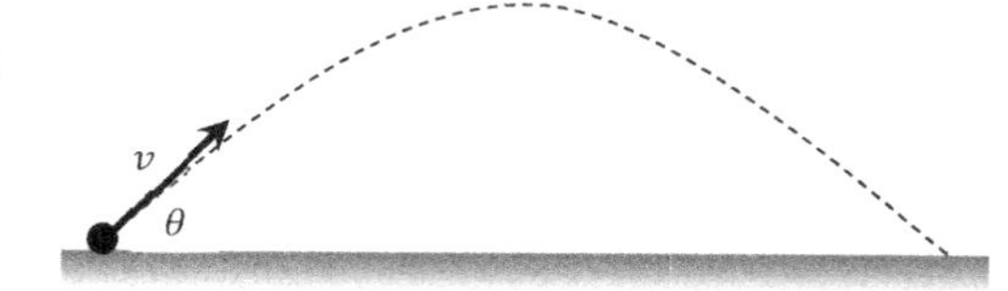

When I say a level ground projectile problem I typically mean the impact and launch locations have equal elevation.

Only in the special case of level ground projectile problems do the following facts apply:

- $t_{\text{to max height}} = \frac{1}{2}t_{\text{enitre flight}}$
- $v_{fy} = -v_{iy}$
- $v_f = v_i$
- Range $= R = \Delta x_{total} = \frac{v_i^2}{g}\sin 2\theta$

Notice the importance of remembering when unit vectors are implied. The second equation discusses scalar components of velocity and the $\hat{\jmath}$ is implied. The third equation discusses impact *speed*; this is a magnitude so no unit vectors are implied.

Heads Up:

In the solutions I use both $y_f = y_i \pm v_i \sin\theta\, t - \frac{g}{2}t^2$ and $\Delta y = \pm v_i \sin\theta\, t + \frac{1}{2}a_y t^2$.
I expect you to be comfortable using $a_y = -g$ and $\Delta y = y_f - y_i$ to go back and forth.

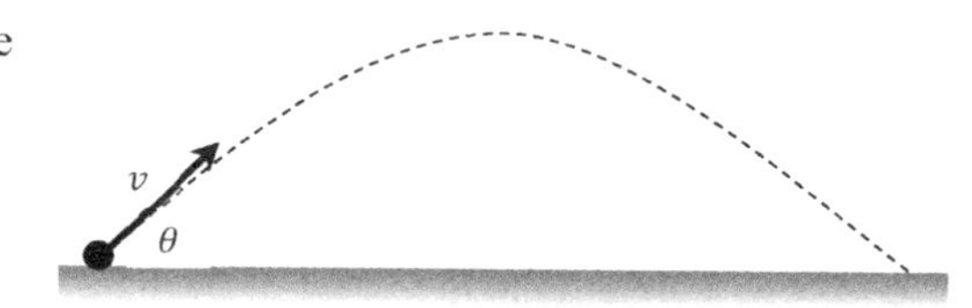

4.8 Consider a level ground problem in which the impact location is at the same elevation as the launch location. Note: we typically assume the initial launch height is negligible if it is small compared to max height. The ball is launched with muzzle speed v and angle θ above the horizontal. Assume air resistance is negligible.

a) Determine the max height.
b) Derive the time to max height.
c) Determine the time of flight.
d) Derive the horizontal range equation. Range is the total horizontal distance for the entire flight.
e) What is the speed at max height?
f) What is the acceleration at max height?

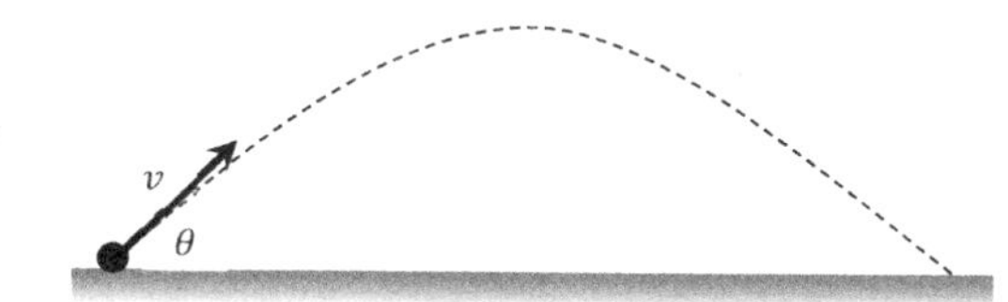

4.9 Requires Calculus: Consider a level ground problem in which the impact location is at the same elevation as the launch location. Note: we typically assume the initial launch height is negligible if it small compared to max height. The ball is launched with muzzle speed v and angle θ above the horizontal. Assume air resistance is negligible.

a) Derive what launch angle gives maximum range.
b) Determine the maximum range.
c) Determine the angle (or angles) that give a horizontal range half of the maximum horizontal range.

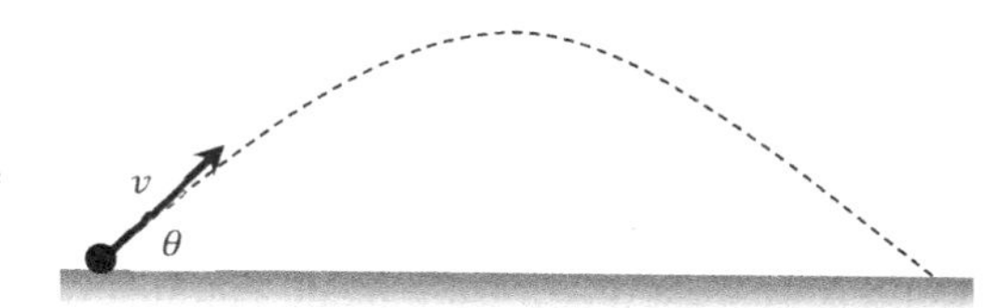

4.10 Consider a level ground problem in which the impact location is at the same elevation as the launch location. Note: we typically assume the initial launch height is negligible if it small compared to max height. The ball is launched with muzzle speed v and angle θ above the horizontal. Assume air resistance is negligible.

a) Determine the ratio of the maximum possible flight time to the flight time for max range (45° launch).
b) Determine the ratio of the maximum possible height to the max height of a 45° launch.

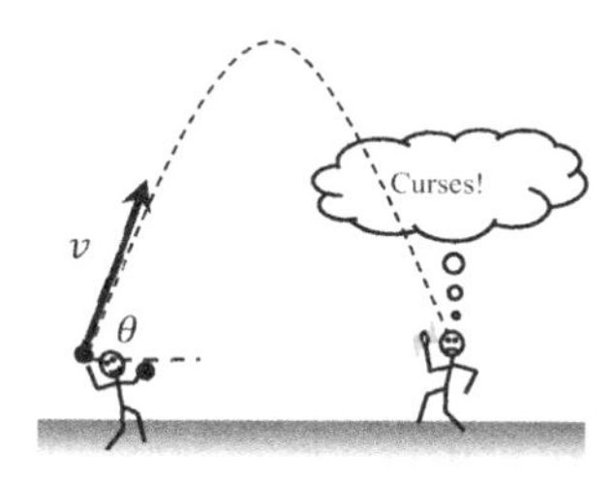

4.11 In a snowball fight Cam throws a ball at 70° above the horizontal with launch speed v. The snowball's trajectory is set up perfectly to hit Dan in the face. Dan is looking up and watching, waiting to knock the snowball down and then counterattack. Cam, mischievous plotter that he is, observes Dan looking up, just as he had planned. He waits a moment and then throws a second snowball with the same speed as the first. The trajectory is set up so the 2nd snowball will hit Dan in the face at exactly the same time as the first snowball. Assume air resistance is negligible. Ignore the heights of the people since they are approximately equal compared to max height. Figure not to scale.

a) What is the distance from Cam to Dan's face in terms of v and g?
b) What angle is required for the second throw?
c) Determine the time delay between the first and second throws.

Note: after eating a face full of snow and ice, Dan retaliates by having his UAV destroy Cam's robot that adheres to ceilings using pressure differentials (discussed much later). While Cam is irritated about losing countless hours of labor, they both laugh about it afterwards and think, "At least we're making more money than our first-year physics instructor!"

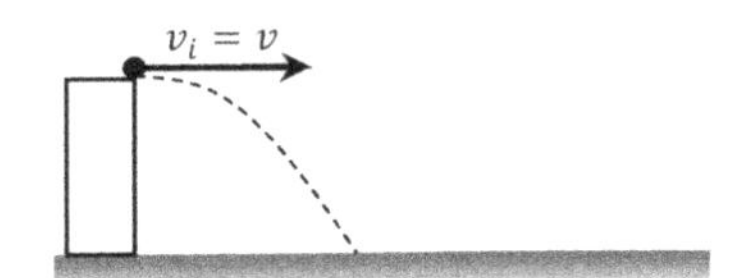

4.12 A ball on top of a building is kicked horizontally with speed $v_i = v$. The building has height h. Assume air resistance is negligible.

a) Determine the speed at impact.
b) Determine the impact angle.
c) Determine the range of the ball.

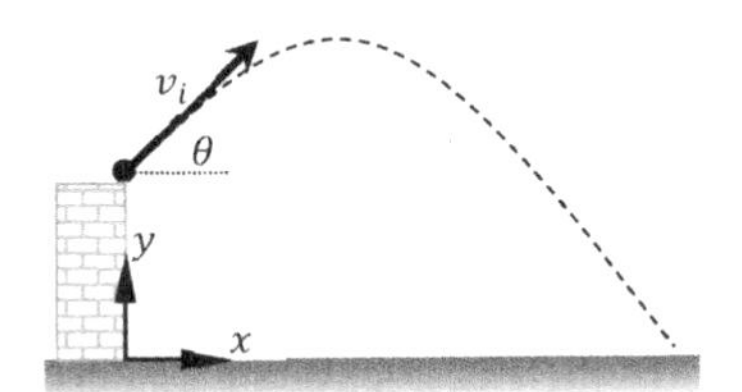

4.13 A ball on top of a 10.0-meter tall building is kicked. Just after the kick the ball has speed 19.6 m/s heading 30.0° above the horizontal. Assume air resistance is negligible. **Do work algebraically. Plug in #'s after solving.**

a) Determine the max height of the ball above the ground.
b) Determine the velocity at max height.
c) Determine the acceleration at max height.
d) Assuming the base of the building is origin, what are the (x, y) coordinates of max height?

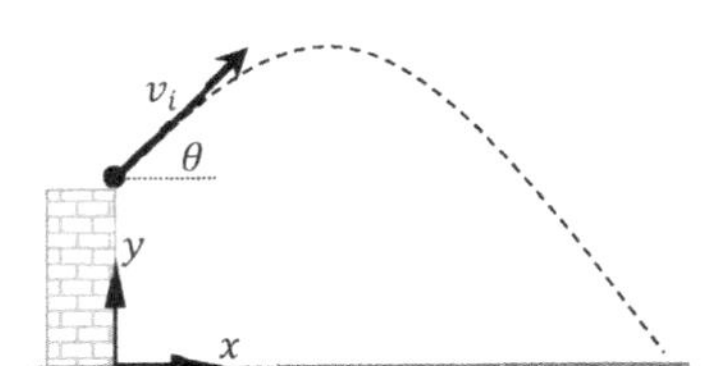

4.14 A ball on top of a 10.0-meter tall building is kicked. Just after the kick the ball has speed 19.6 m/s heading 30.0° above the horizontal. Assume air resistance is negligible. **Do work algebraically. Plug in #'s after solving.**

a) Determine the range of the projectile.
b) Determine the impact velocity. Express your result in polar form (magnitude and direction). Sketch the final velocity as an arrow and indicate an angle.

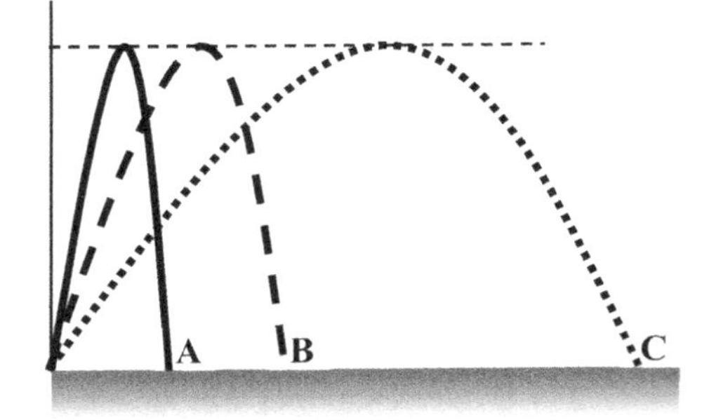

4.15 The figure at right shows the trajectories of three different balls that were launched by students. The trajectories are labeled **A, B, & C** to aid in discussions. For this problem, ignore any effects of air resistance. Figure not to scale.

a) Rank the initial *vertical velocity components* of each of the three balls from greatest to smallest. Clearly indicate any ties.
b) Rank the initial *speeds* (greatest to smallest). Clearly indicate any ties.
c) Rank the *time of flight* of each ball (greatest to smallest). Clearly indicate any ties.

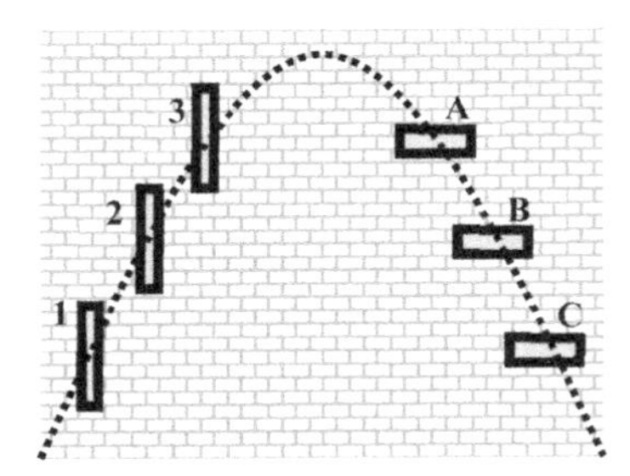

4.16 A projectile passes by several funky signs on a building. Some are tall and skinny. Others are wide and short. The path of the trajectory as the projectile passes the signs is shown in the figure.

a) Rank the average speed while crossing each sign for signs **1**, **2**, & **3**.
b) Rank the time spent passing each sign for sign **1**, **2**, & **3**.
c) Rank the average speed while crossing each sign for sign **A**, **B**, & **C**.
d) Rank the time spent passing each sign for sign **A**, **B**, & **C**.

4.17 A projectile is launched from the top of a cliff with initial speed of $12.3\frac{\text{m}}{\text{s}}$ at an angle of 60.0°. The ball impacts the ground with twice the speed. Determine the height of the cliff. Assume air resistance and the height of the person throwing the ball are negligible.

4.18 A watermelon is thrown beneath the surface of a swimming pool. It impacts the surface of the water moving with speed $v = 6.00\frac{\text{m}}{\text{s}}$ at an angle of $\theta = 30.0°$ below the water's surface. While underwater, the forces acting on the melon are quite complicated. The forces acting on the ball are drag, buoyant force, and gravity. To simplify the problem and make it doable, let us assume the net effect of all these forces is to cause acceleration in the x-direction and the y-direction. At the end of the problem we can discuss how reasonable we think these assumptions are. For now, assume $a_x = -2.00\frac{\text{m}}{\text{s}^2}$ and $a_y = 4.00\frac{\text{m}}{\text{s}^2}$ based on the coordinates shown in the picture. Notice the negative value for a_x implies the melon should slow down as it moves sideways. The positive value for a_y implies the melon should slow down until it reaches some max depth then speed up as it returns to the surface.

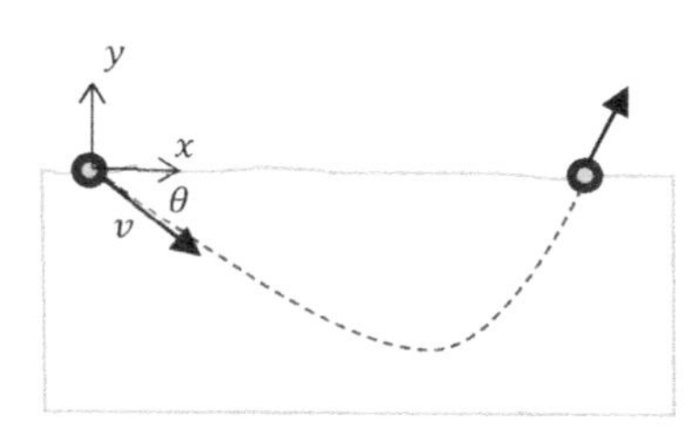

a) Determine the final speed and heading of the melon as it resurfaces.
b) Determine the time to resurface.
c) Determine the horizontal distance traveled while the melon is underwater.
d) Determine the max depth.
e) Do you think our assumed values of the acceleration are reasonable? Explain.

4.19 At time $t = 0$ a drone is at the origin travelling with speed v_0 to the right. The acceleration of the drone is given by $\vec{a} = -\frac{g}{2}\hat{\imath} + 2g\hat{\jmath}$ where g is the standard constant (*magnitude* of freefall acceleration). To be clear, this acceleration accounts for all forces on the drone including drag, gravity, and thrust.

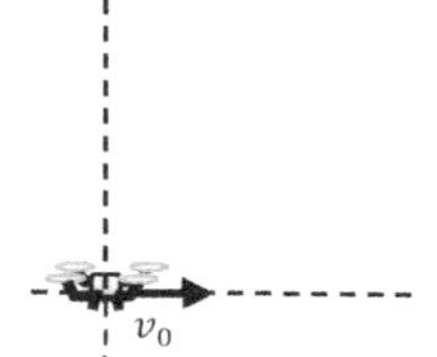

a) At what time will the drone reverse direction horizontally?
b) How far is the drone from the origin when it reverses direction horizontally? Notice I said how far, not how far horizontally…
c) What is the speed of the drone when it passes above its initial position?
d) What is the direction of motion of the drone when it passes above its initial position?

4.20 You practice throwing rocks for several hours at a gong. One time you try it blindfolded. You throw a ball at known angle θ above the horizontal. You release it distance h above the ground. You hear the ball hit the gong hanging at the corner of a building. The gong is horizontal distance x from the launch point. The building has height H.

a) Determine the launch speed of the ball.
b) Determine the final speed of the ball.
c) Explain how you cannot distinguish between two possible final velocities of the ball.

4.21 An eagle is grasping an extremely aerodynamic taco as it flies downwards in a steep dive as shown in the figure. The eagle's speed is 60.0 m/s (~130 mph) when it is 30.0 m above the surface of a lake. At this instant, the eagle spies a fish near the surface of the lake and releases her grasp on the taco. The taco hits the lake after a total displacement of 33.0 m. Determine the angle of descent at the instant the eagle releases the taco. Assume air resistance is negligible.

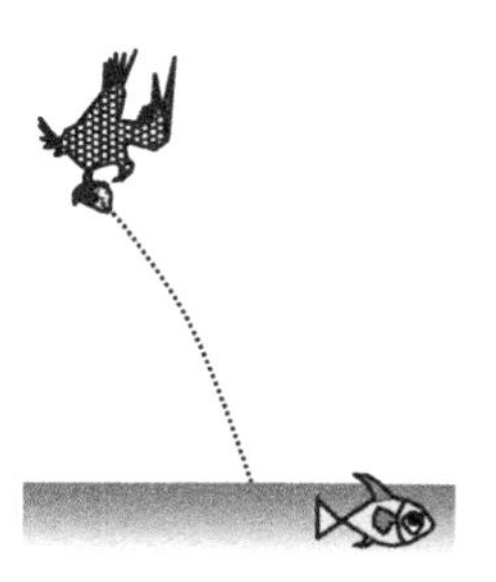

65.4° below the horizontal is *close* but not correct…

4.22 A dart is launched at a tiny monkey as shown in the figure. As the dart leaves the gun it makes a sound. The monkey hears this sound and reacts almost instantaneously by letting go of the branch.

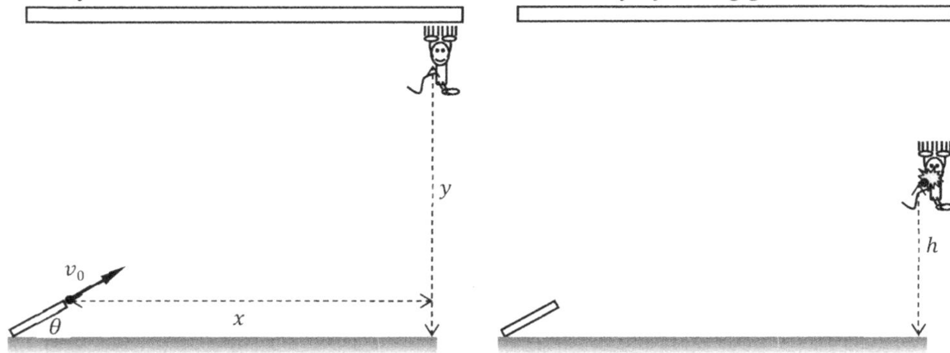

a) Guess the direction one should aim the dart to hit the monkey. Should it be above, below, or directly at the monkey? What if the gun has small or large muzzle speed? How will muzzle speed affect things? Again, assume the monkey lets go of the branch at the instant that the dart leaves the gun.

b) Determine the impact height h in terms of y, g, and the time of flight t. Hint: monkey around with it.

c) Determine the launch angle θ required to cause impact. Answer in terms of only x and y. Compare your result to your guess in part *a*!

d) Is your previous answer valid when the gun is even with the monkey? What about a gun initially above the monkey? What if the muzzle speed is fast or slow?

e) **Variation:** Assume in the problem that x and y are given. In theory you would also know θ since you could figure it out from x and y. In addition you decide you want to hit the monkey when it has fallen to height h above the ground. Determine the launch speed required in terms of x, y, h, and g. Show you get

$$v_0 = \sqrt{\frac{g(x^2 + y^2)}{2(y - h)}}$$

Does the formula make sense if $y = h$? Does it make sense when $h = 0$?

4.23 In reality air resistance is a huge factor in this problem but ignore it for now so the problem is doable.

A ski jumper launches off a ramp with a slight incline of θ as shown in the figure. The ski jumper's initial velocity is v. The ski jumper travels a total distance L down the slope. The slope has an angle of ϕ above the horizontal as shown in the figure below.

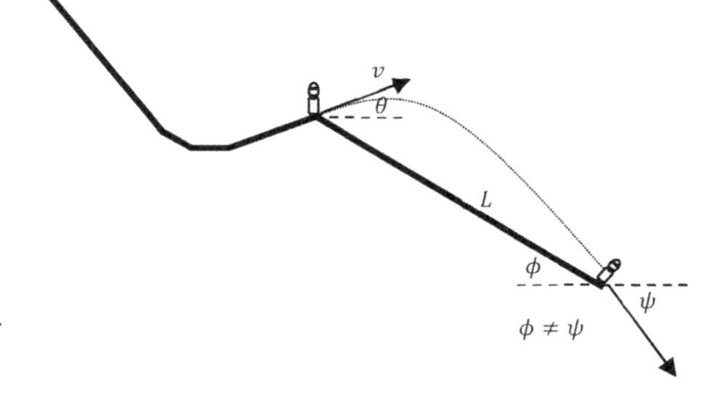

People like big jumps. As a result, some ski jump slope engineer wants to make sure the angle at the end of the ramp gives the longest jumps possible.

a) Show the downslope range is $L = \frac{2v^2 \cos\theta}{g \cos^2\phi} \sin(\theta + \phi)$.

b) **Requires Calculus:** Determine the angle θ that maximizes downslope range.

c) Verify both previous results make sense in the limit $\phi = 0°$.

Some potentially useful trig identities are

$$\sin(A \pm B) = \sin A \cos B \pm \cos A \sin B$$
$$\cos(A \pm B) = \cos A \cos B \mp \sin A \sin B$$

4.24 Requires Calculus: You are told that the position vector of a particle, as a function of time, is given by the vector expression

$$\vec{r} = (6 - 2t)\hat{\imath} + 3t^2\hat{\jmath} - 4\hat{k}$$

The $+x$-direction corresponds to right, $+y$-direction is up, and $+z$-direction is out of the paper. Assume the units of position are m.

a) Assuming t is time, what units must be on each number to be dimensionally correct.
b) What is the initial position?
c) What is the initial speed? What direction is the object initially travelling?
d) What is the initial acceleration (mag and direction)? Is this a constant acceleration problem?
e) Write down the speed v_f as a function of time. Are there any times when the speed is zero?
f) It is easier to determine displacement as opposed to distance traveled for an arbitrary time t. Explain why.
g) How would one figure out the distance traveled? Can you set it up?

4.25 Requires Calculus: Suppose initial position of a particle is at the point (1, 2, 4). The velocity is given by

$$\vec{v} = 2t\hat{\imath} - (5 - 3t^2)\hat{\jmath} + \frac{4}{(t+1)^2}\hat{k}$$

The $+x$-direction corresponds to right, $+y$-direction is up, and $+z$-direction is out of the paper. Assume the units of velocity are m/s. Use this to determine the units on all constants and answers below.

a) Determine the position as a function of time in Cartesian form...$\vec{r}(t) = x(t)\hat{\imath} + y(t)\hat{\jmath} + z(t)\hat{k}$.
b) At $t = 1$ sec, how far is the particle from its starting position?
c) Determine the acceleration as a function of time in Cartesian form.
d) What is the speed as a function of time?
e) Does the particle ever reach a max or minimum height in the y-direction? If no, explain how you can tell from the equation. If yes, determine the (x, y, z) coordinates at the max or min height. Clearly indicate if the particle is at a max or a min.
f) What angle does velocity make with the positive y-axis at $t = 1$ sec?

Relative velocity

Consider two particles **A** & **B** about to collide as shown below. The figure on the left shows both particles moving relative to the stationary earth. The figure on the right views the motion from **B**'s perspective. **B** is at rest while **A** moves towards it.

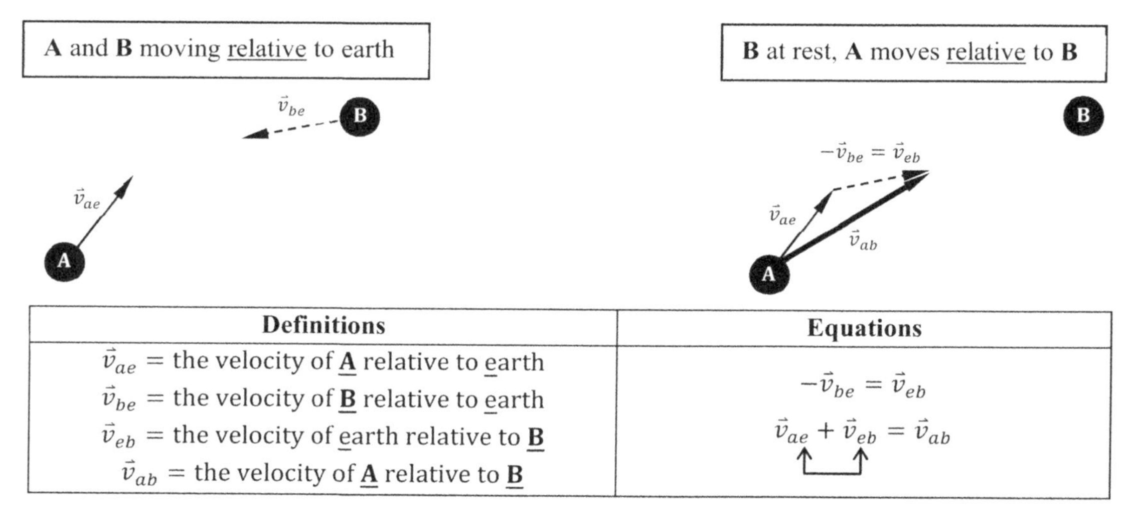

Definitions	Equations
$\vec{v}_{ae}$ = the velocity of **A** relative to earth $\vec{v}_{be}$ = the velocity of **B** relative to earth $\vec{v}_{eb}$ = the velocity of earth relative to **B** $\vec{v}_{ab}$ = the velocity of **A** relative to **B**	$-\vec{v}_{be} = \vec{v}_{eb}$ $\vec{v}_{ae} + \vec{v}_{eb} = \vec{v}_{ab}$

These definitions and equations are meaningless without practice. While the problems appear to be simple vector addition, history has shown students often get confused.

4.26 A spy drone hovers above a busy marketplace in Santa Maria. The drone observes student **A** walking due east with speed 2.00 m/s relative to earth. Simultaneously, student **B** hustles due west with speed 3.00 m/s relative to the earth.

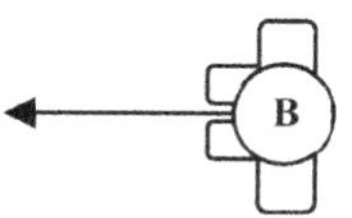

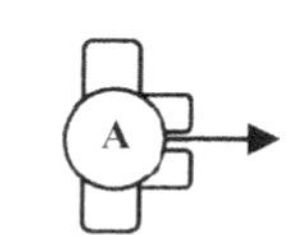

a) Determine the velocity of **B** relative to **A**.

b) Sketch the picture from **A**'s perspective. To get a feel for this, imagine the drone is hovering above student **A**. Which way and how fast does student **B** move?

4.27 This time student **A** walks due east with speed 2.00 m/s relative to earth. Student **B** walks due south with speed 3.00 m/s relative to the earth.

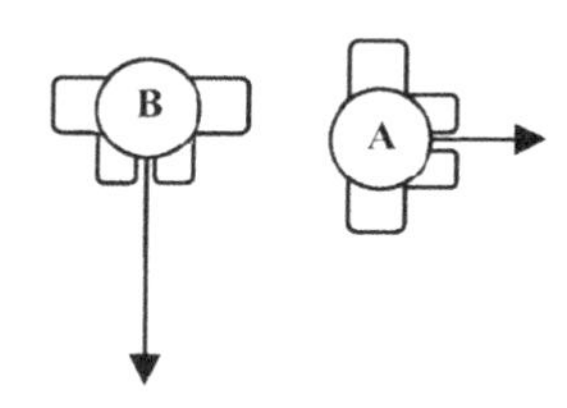

a) Determine the velocity of **B** relative to **A**.

b) Sketch the picture from **A**'s perspective. To get a feel for this, imagine the drone is hovering above student **A**. Which way and how fast does student **B** move?

4.28 This time student **A** walks with speed 4.00 m/s directed 30.0° north of east relative to earth. The drone, following student **A**, records student **B** moving with speed 8.00 m/s directed 50.0° south of west. How fast and in what direction is student **B** moving *relative to the earth*?

4.29 In still water a boat can travel at 10 m/s. This boat is travelling up and downstream in a river flowing with speed 2 m/s (relative to the earth). Assume the river flows due east ($\hat{\imath}$).

a) Determine the upstream and downstream velocities of the boat relative to the earth. I know the answers are pretty obvious but please try to use this as an opportunity to use the notation mentioned above.

b) Suppose the boat travels upstream (due west) for 10 s then immediately turns around and travels downstream (due east) for 10 s. Determine the displacement, distance traveled, average velocity, and average speed of the boat relative to the earth.

c) Now the boat travels upstream for 96 m then downstream for 96 m. Determine the total travel time. Also determine the displacement, distance traveled, average velocity, and average speed of the boat relative to the earth.

4.30 In still water a boat can travel at 10.0 m/s. This boat is to cross a river which is running at 2.00 m/s to the east. Answer the following questions:

a) At what angle θ should the boat aim to travel straight across the river to the opposite shore?

b) Assume the boat travels due north across the flowing river. If the shore is 50.0 m away, how long will it take to get there?

c) On the way back to the first shore, the boat aims so that it crosses the river as quickly as possible. In which direction should the boat aim & how much faster is this crossing than the previous?

Opposite shore
Direction boat travels
Direction to aim?
θ
River flows 2 m/s
Starting point

4.31 At time $t = 0$ a bat is distance d due south of the origin flying at speed v heading 60° east of north. At the same instant, a moth located distance d due east of the origin is flying with speed $v/2$ due west. Assume the speeds listed are measured by a stationary observer.

a) Determine the speed of moth relative to the bat.

b) Determine an equation for the position of the moth relative to the bat as a function of time.

c) **Mega-Challenge:** At $t = 0$ the bat begins to accelerate with constant rate a. What direction must the bat accelerate to intercept the moth? Where will they intersect and at what time?

Note: it is interesting to read about the sonar jamming techniques employed by moths to counter bat attacks.

4.31½ A train is traveling to the right with a constant speed of $20.0\,\frac{\text{m}}{\text{s}}$ *relative to the earth.* A baseball player throws a ball with an initial speed of $40.0\,\frac{\text{m}}{\text{s}}$ *relative to the train.* At what angle should the player aim to throw the ball if she wants the ball to leave her hand with a launch angle of 30.0° *relative to the stationary ground.*

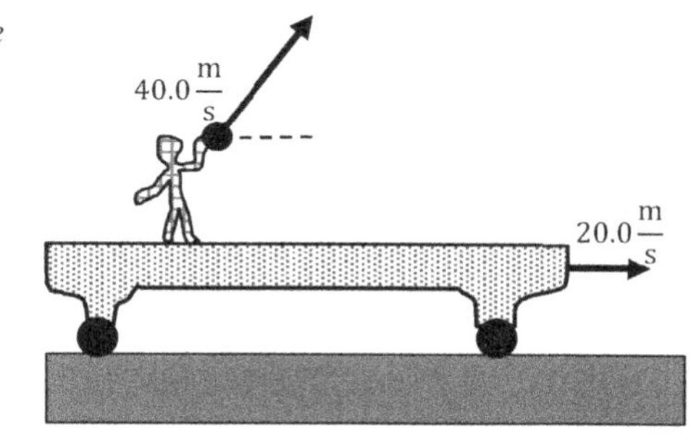

4.32 Numerical work. Most of the 2D motion problems we've discussed had no acceleration in the x-direction. Let us now consider an astronaut in space with thrusters for both the x- and y-directions. This astronaut considers her space ship as the origin so her initial position could be non-zero for either direction. Furthermore, she might have initial speed in either direction as well as acceleration in either direction. Her horizontal and vertical positions are determined by

$$x(t) = x_i + v_{xi}t + \frac{1}{2}a_x t^2$$

$$y(t) = y_i + v_{yi}t + \frac{1}{2}a_y t^2$$

while her distance from the spaceship is given by

$$r(t) = \sqrt{[x(t)]^2 + [y(t)]^2}$$

The velocity components are given by

$$v_x(t) = v_{xi} + a_x t$$
$$v_y(t) = v_{yi} + a_y t$$

while her speed is given by

$$v(t) = \sqrt{[v_x(t)]^2 + [v_y(t)]^2}$$

Use these equations to generate a table like the one shown below. The top six cells are constants referenced in the table below. Use your tables to plot y vs x, r vs t, and v vs t. I created these and they are shown below. Try changing the constants (your initial parameters) and see the weird plots you can make (shown at right). I thought the initial parameters chosen below made for interesting plots.

a) Is the first curve a function? Does it need to be? Is the derviative of the first curve useful for determining the speed, the velocity, or something else?

b) What parameter needs to be set to zero to get standard projectile problems?

c) **Challenge:** Under what conditions is the motion of the astronaut 1-dimensional?

Imagine how scary your first untethered space walk would be. We haven't even accounted for the third dimension or rotation yet and this already starts to look fairly tricky…it appears to be no joke getting back to the ship.

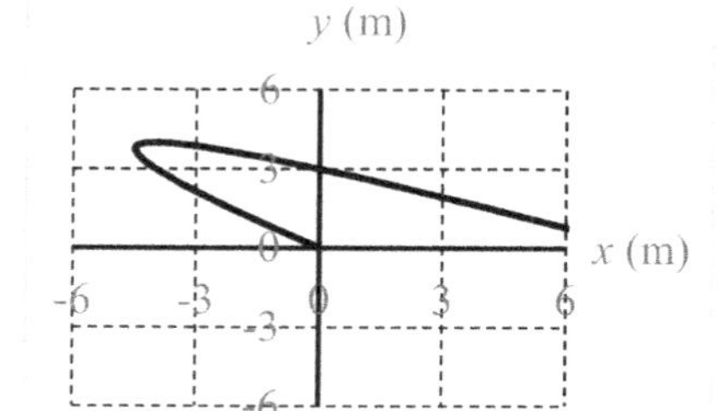

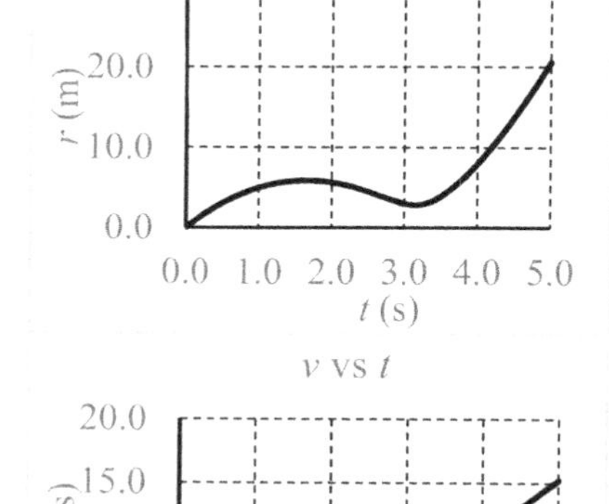

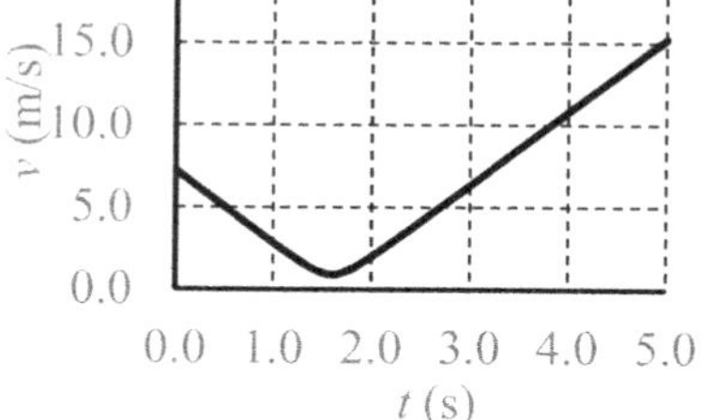

x_i (m)	v_{xi} (m/s)	a_x (m/s²)
0	-6	4

y_i (m)	v_{yi} (m/s)	a_y (m/s²)
0	4	-2

t (s)	x (m)	y (m)	v_x (m/s)	v_y (m/s)	r (m)	v (m/s)
0.0	0	0	-6	4	0.0	7.2
0.2	-1.12	0.76	-5.2	3.6	1.4	6.3
0.4	-2.08	1.44	-4.4	3.2	2.5	5.4
0.6	-2.88	2.04	-3.6	2.8	3.5	4.6
0.8	-3.52	2.56	-2.8	2.4	4.4	3.7
1.0	-4	3	-2	2	5.0	2.8
…	…	…	…	…	…	…
5.0	20	-5	14	-6	20.6	15.2

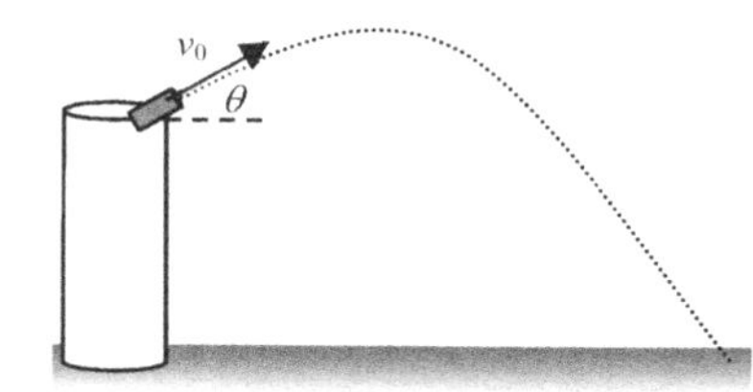

4.33 Numerical work. Consider a projectile with initial height h, launch speed v_0, and launch angle θ. It is common to assume that the optimum angle for maximum range is 45°. This is not true, even for our simple model without air resistance. The goal of this problem is to determine angle of the maximum range of a projectile when the launch position is above the impact position.

a) Determine an algebraic expression for the impact speed in the y-direction in terms of the givens and g. Hint: double check the sign of your final answer.

b) Determine an algebraic expression for the time of flight in terms of the givens and g.

c) Determine and algebraic expression for the range of the projectile in terms of the givens and g. Check the units on your result. Also check that it reduces to the level ground range equation when $h = 0$.

d) Set $v_0 = 20\frac{\text{m}}{\text{s}}$ and $h = 10$ m. Tabulate and plot x vs θ. Use your plot (or table) to show the max range occurs when $\theta = 39°$. Compare this to the simulation questions 4.

e) **Challenge:** Hopefully you set up a table to use equations and reference the first row of constants. If so, it should be easy to change the initial speed or initial height. Use your model to determine if the optimum launch angle (θ_{max}) depends on the v_0 and/or g. For reference, we know that for a level ground problem θ_{max} is always 45° independent of v_0 and g. For reference, some other values of g are listed in the table.

Location	g (m/s²)
Moon	1.63
Mars	3.75
Earth	9.8
Jupiter	26.0

4.34 A ball is shot with initial speed v_0 and launch angle θ towards a parabolic structure. Using the launch position as the origin, the structure's parabolic equation is $y = \alpha x^2$. Determine the time to impact.

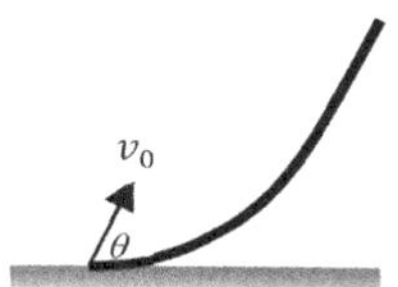

4.35 A hockey puck is hit horizontally at the top of hemispherical structure. Just after the puck is hit, the puck has initial speed v_0. The structure has radius R.

a) Determine distance between point where puck lands and edge of hemisphere.

b) **Challenge Requires Calculus:** What *minimum speed* allows the puck to reach the ground without ever touching the edge of the hemisphere?

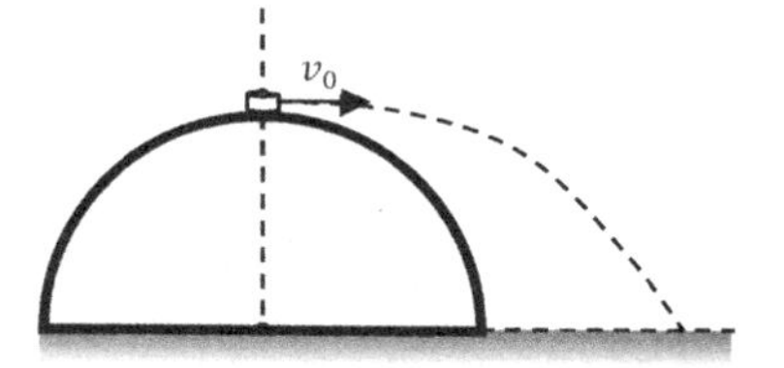

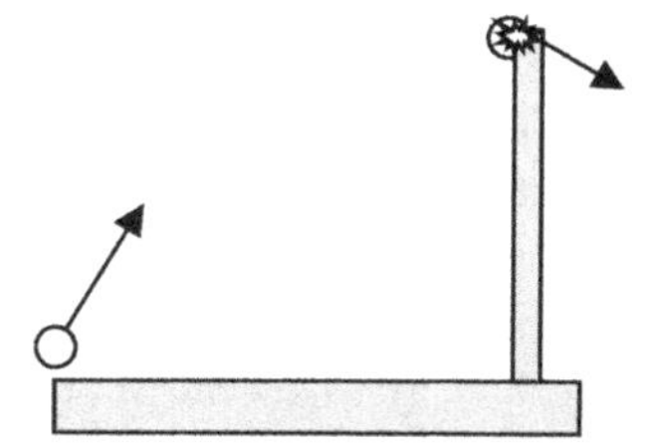

4.36 A ball is launched from negligible initial height. The ball travels distance x horizontally and impacts a the top of a post of height H. The ball was launched with known initial speed v. Determine all possible launch angles θ. If you have trouble, this should be similar to **4.19** and **4.20**.

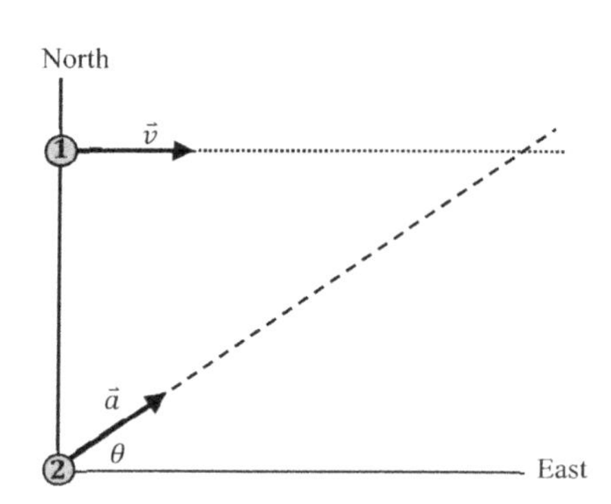

4.37 A top view of two particles is shown. Particle **1** is initially 60.0 m due north of particle **2** and travels due east with constant speed 6.00 m/s as shown in the figure. At the instant particle **1** passes the north axis, particle **2** accelerates from rest with magnitude 0.800 m/s^2. What angle θ causes a collision?
To be clear this is not a projectile problem under the influence of gravity. Think of it more like having a bird's eye view of two cars coming into an intersection.

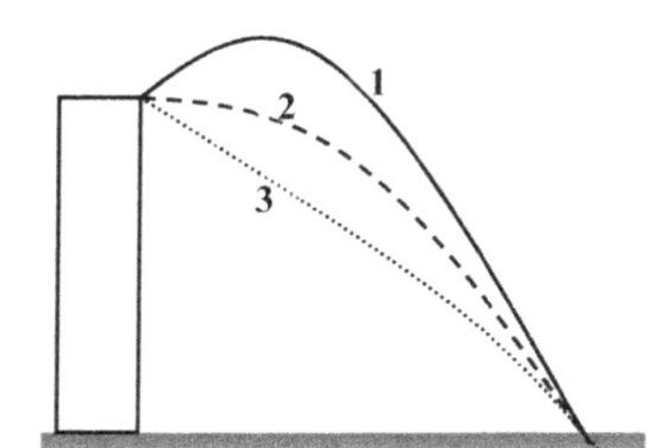

4.38 Challenge: Three rocks are thrown from the top of a building at different angles. Assume rock 1 has a launch angle of 30° *above* the horizontal, rock 2 is thrown horizontally, and rock 3 is launched 30° *below* the horizontal. **The three rocks all impact the same location.** The trajectory of each rock is shown in the figure. Air resistance is negligible. Figure not to scale.

a) Rank the horizontal components of velocity of the rocks.
b) Rank the time of flight for each rock. Clearly indicate any ties.
c) Rank the launch speeds of the rocks.
d) Rank the impact speeds.

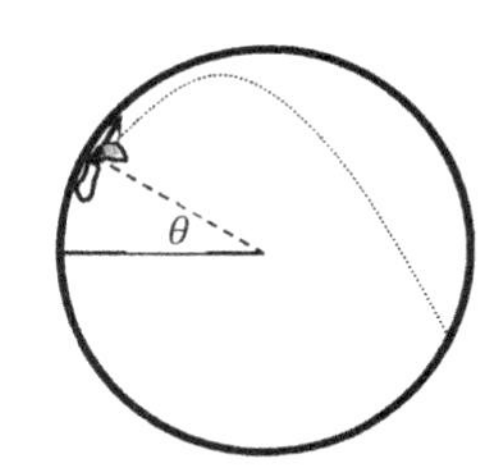

4.39 Challenge: A small sock spins in a large dryer with radius $R = 1.00$ m. Think…that's a HUGE dryer! The sock loses contact with the walls of the machine and becomes a projectile when $\theta = 60.0°$ travelling with speed $v = 2.913\frac{\text{m}}{\text{s}}$. At what (x, y) coordinates will the ball impact the washing machine? Hint: it is often wise to put your coordinate system at the center of a circle so you can use the equations for circles that you have memorized from math classes.

Forces

Forces are vectors with magnitudes and directions. Many, many times we will be trying to determine the magnitude of a force separately from its direction. A list of common forces is shown below. I have included the symbol I plan to use for each type of force. The units for force are Newtons (N) where $1\text{ N} = 1\ \frac{\text{kg}\cdot\text{m}}{\text{s}^2}$

An easy way to remember the units is to think "The units of force are the units of mg."

Type of Force	Symbol for Magnitude of Force	Comments
Normal	n	Acts at the interface between two surfaces. Acts perpendicular to each surface. Acts radially for spherical/cylindrical surfaces. The electrons on the atoms of one surface are essentially repulsing the electrons on the other surface.
Tension	T	Typically the force in a cable or string. Unless otherwise specified, we often assume the cables/strings are massless and inextensible.
Gravity	mg or $\frac{Gm_1m_2}{r_{12}^2}$	mg applies near the Earth's surface. The other equation applies for astronomical size problems (satellite orbits, etc). Remember the following • Gravity is a force • g is NOT gravity… • g is the magnitude of the acceleration due to gravity (in freefall)
Friction	f	Friction opposes the direction of motion. Sometimes the direction of motion is not the direction of acceleration. Said another way $\vec{f}$ points opposite $\vec{v}$; $\vec{f}$ can point the same way **or** the opposite way of $\vec{a}$.
Push	Usually F	A lot of problems will say "A force with magnitude F is applied…". Typically this is a person pushing/pulling on the block.
Pull	Usually F	
Spring	$F_{spring} = kx$	Watch out! $\vec{F}_{spring} = -kx\hat{\imath}$ but $F_{spring} = kx$ See the difference?
Electric	$\frac{kq_1q_2}{r_{12}^2}$	The equation shown is from something called Coulomb's law. More on this later. For now, notice that the astronomical gravity equation and this equation are very similar. Maybe you should actually pay attention in Chapter 13 so you suffer less in a subsequent course…
Magnetic	…	Wait until you understand electricity.
Strong	…	In the nucleus. Think why don't protons in a nucleus fly apart? I thought like charges repel? This puppy comes into play.
Weak	…	Another force important in understanding the nuclei of atoms. Note: the four fundamental forces of physics are strong, electromagnetic, weak, and gravitation (ranked from strongest to weakest).

Notice:

- In this class usually lower case t is *time* while, for now, upper case T is *tension.*
- Lower case n is *normal force* while upper case N is Newtons (the *units* of force)
- Lower case f is *frictional force* while upper case F is typically an applied push/pull force.
- The opposite of a push is a pull.
- Typically we try to draw force pictures so the magnitudes are positive. For instance, we like to think of normal forces as positive. A negative magnitude for a normal force implies the surface is pulling instead of pushing. Similarly, a negative tension magnitude would imply you are pushing on something with a string…seems a bit strange to me. That said, there will be times we may actually want to use this style…

Newton's 1st Law

- An object in motion stays in motion unless acted on by non-zero net external force.
- An object at rest stays at rest unless acted on by non-zero net external force.
- An object moves with constant *velocity* (magnitude *and* direction) unless acted on by non-zero net force.
- If $\Sigma\vec{F} = 0$ then $\vec{a} = 0$.

Newton's 2nd Law

- Acceleration of object is directly proportional to net force and inversely proportional to mass.
- $\Sigma\vec{F} = m\vec{a}$: if net force is non-zero, the object accelerates.
- $\vec{a} = \frac{\Sigma\vec{F}}{m}$: for a given net force, the larger the mass the smaller the acceleration.
- Mass = inertia. The more mass (inertia) you have, the more you resist changes in velocity.

5.1 A tale of two strings

a) A ball of large mass is tied to the ceiling with a string. A second string is tied to the bottom of the ball. A student gives a quick, hard jerk on the bottom string. Which string breaks? Explain.

b) Suppose the teacher pulls on the bottom string differently. This time, she gradually increases the tension in the bottom string until a string breaks. Which string breaks? Explain.

Broomstick on two wineglasses

A fun physics demo involves a broomstick resting on two wine glasses as shown in the figure. We usually use a 1" dowel rod instead of a broomstick. A headless pin is placed in each end of the dowel. By placing the pins slightly off center, the dowel will rest on the glasses without rolling off. Someone whacks the center of the dowel with a metal pipe. The same demo was supposedly done by hanging the dowel on loops of thread attached to circus performers' ears…

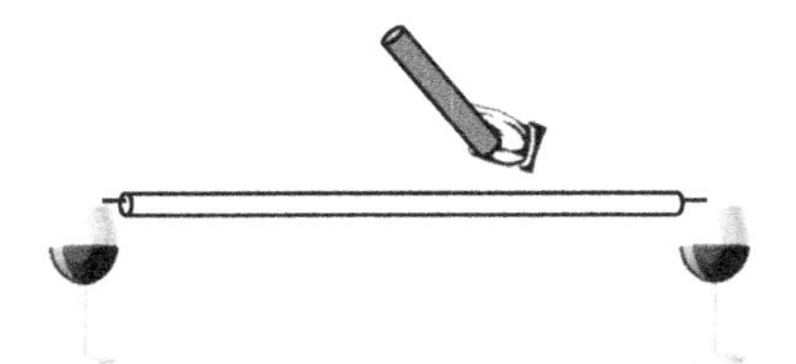

5.2 Table Cloth Trick

Suppose we have a very smooth tablecloth. The tablecloth has no large seams on the edge. The tablecloth sits at rest on a table that is also very smooth. On top of the table cloth, sits block of mass *M*. Starting from rest, the tablecloth is jerked very quickly so that it moves horizontally out from underneath the block. Which is easier: using an empty plastic plate or a ceramic plate full of food? What about a full vs empty wine glass? Explain.

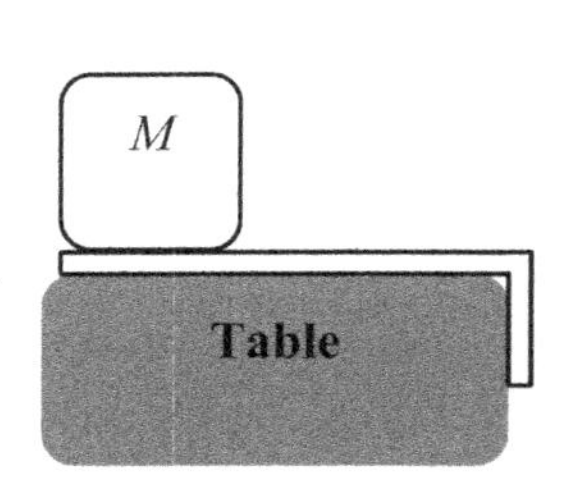

Inertia or Impulse?

The previous demos demonstrate the property of inertia but inertia alone is insufficient to explain them. Each of the demos has a step that requires applying the force rapidly. This relates to impulse discussed in Chapter 9.

- If you pull the tablecloth quickly the frictional force is applied for a tiny amount of time. The block accelerates for a tiny amount of time. The block gains a tiny bit of speed and moves a tiny bit but is quickly stopped once it lands on the table. It ends up with negligible horizontal displacement.
- In the tale of two strings, jerking the bottom string applies force for a very short duration. As a result, the large mass doesn't have significant time to accelerate; it won't move enough to break the upper string.
- In the broomstick demo, the ends of a 2-meter stick move downwards approximately 1 mm during the breaking process. The pins act as cushions during that slight compression on the wine glasses. The center of the stick breaks before the collision shock wave propagates to the ends of the stick.

5.3 An elevator is accelerating downwards with magnitude a as shown in the figure. A mass hangs from a scale connected to the ceiling. The scale measures the tension in the bottom string.

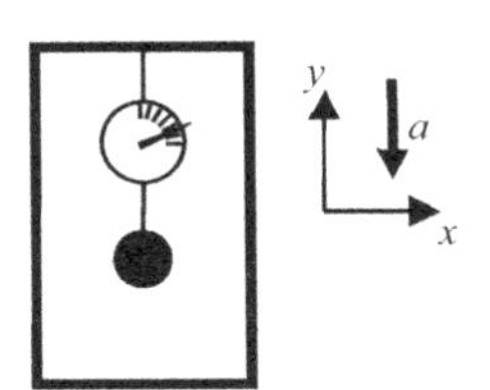

a) Write down the proper force equation (for the hanging mass) associated with Newton's 2nd law ($\sum \vec{F} = m\vec{a}$). Solve the equation for the scale reading.

b) Under what circumstances does the scale read the true weight of the mass?

5.4 An elevator is accelerating downwards with magnitude a as shown in the figure. A person stands on a scale. **Notice the coordinate system has flipped!**

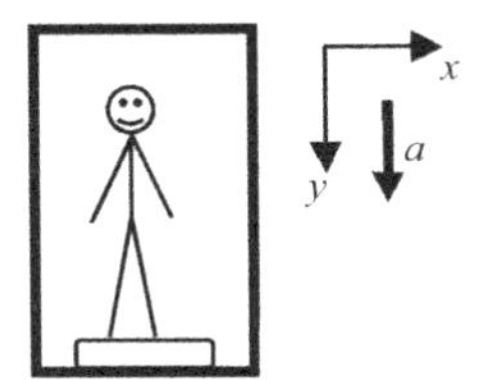

a) Write down the proper force equation for the person associated with Newton's 2nd law ($\sum \vec{F} = m\vec{a}$). Solve the equation for the scale reading.

b) Under what circumstances does the scale read the true weight of the person?

5.5 Two 1.0 kg masses are suspended on strings as shown in the figure. A scale connects the two strings above the table. Assuming g =10 m/s² for this problem, what does the scale read?

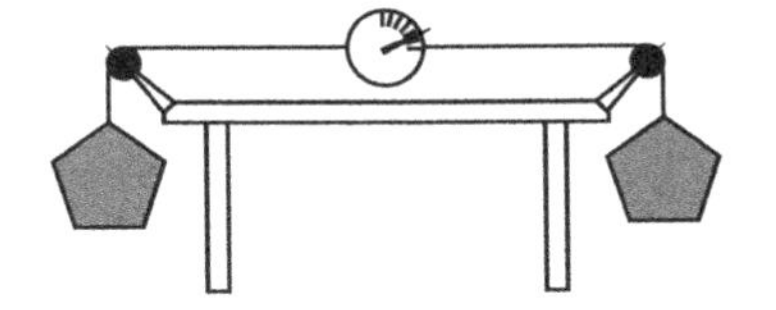

5 N 10 N 20 N

5.6 A mass m is in equilibrium as shown in the figure. The mass is held up by a system of strings as shown. Each string is tied to the washer of negligible mass at the center.

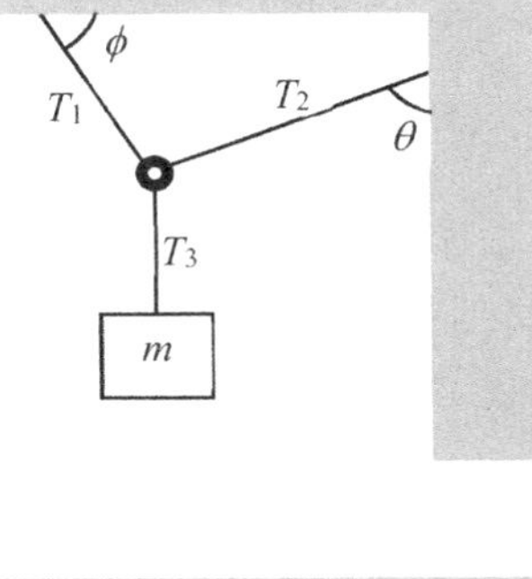
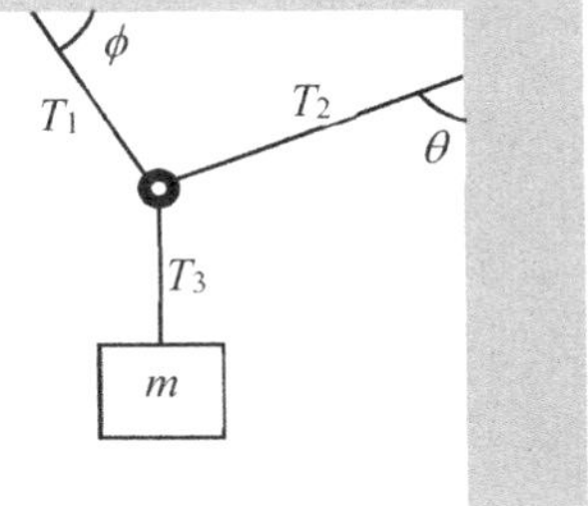

a) Under what circumstances does $T_3 = mg$? Hint: mention a.

b) Draw an FBD for the washer. You can ignore mg of the washer compared to the T's.

c) Write the force equations for the x and y directions based on your coordinate system.

d) Combine your force equations to eliminate the variable T_2 and solve for the variable θ. Your final answer should be in terms of T_1, ϕ, m and g.

e) Now, instead of solving for θ, combine your original force equations to eliminate θ and solve for T_2. Your final answer should be in terms of T_1, ϕ, m and g.

5.7 What if only one string is used and the mass is hooked onto it? Assume the strings are connected to the wall/ceiling with bolts. Assume the mass is free to slide on the string and does so with negligible friction. When this is the case, the angles shown are equal.

a) Draw the FBD of the hanging mass.

b) Determine the tension (magnitude) in the string in terms of m, g & ϕ.

c) What is the tension in the string when $\phi = 90°$? Does your result make sense?

d) What angles cause tension in the cable to exceed the weight of the hanging mass?

e) For fun: contemplate how this relates to design challenges for clotheslines and/or orthodontia (i.e. braces for teeth).

5.8 A zombie pulls a suitcase of mass m to the right in an elevator as shown in the figure. The zombie pulls with tension T at angle θ above the horizontal. In reality the suitcase has small wheels but we may accurately model it as sliding without friction. The elevator accelerates upward with magnitude a_e.

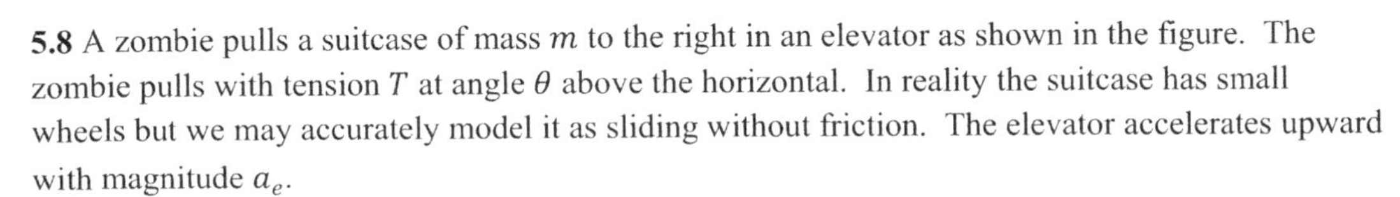
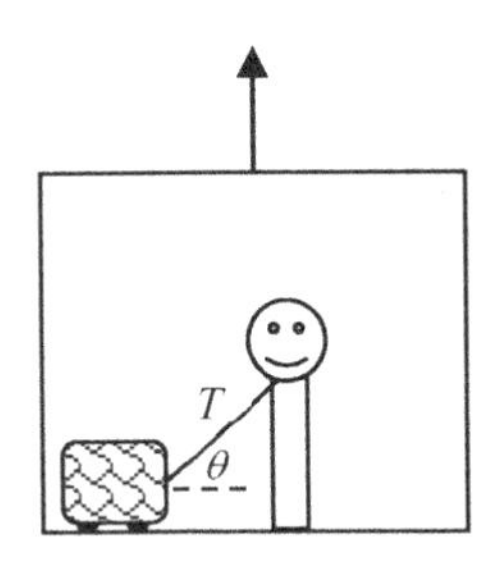

a) Draw an FBD and list the force equations for suitcase.

b) Determine normal force (magnitude) on the suitcase in terms of m, g, T, a_e, and θ.

c) Determine the acceleration (vector) of the suitcase in terms of the givens.

5.9 Now the zombie pulls a suitcase up a ramp with unknown tension $\vec{T}$ at angle ϕ above the plane of the ramp. The zombie and suitcase move parallel to the ramp with *constant speed* v. Friction is negligible. The mass of suitcase is m. The ramp is angled at θ above the horizontal.

a) Draw an FBD and list the force equations for suitcase.
b) Determine the tension (magnitude) in terms of g, m, and the angles.
c) Determine the magnitude of the normal force on suitcase in terms of g, m, and the angles.

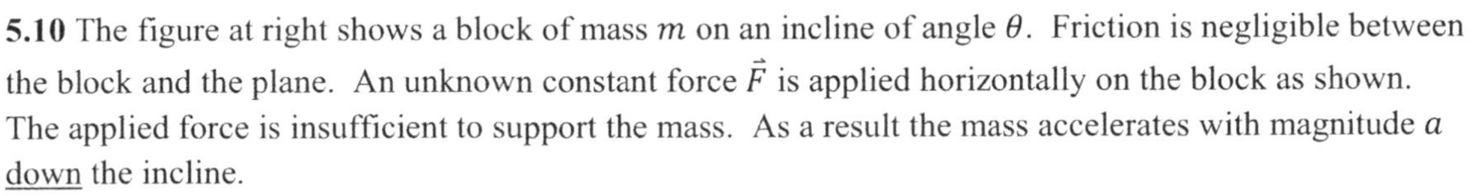

5.10 The figure at right shows a block of mass m on an incline of angle θ. Friction is negligible between the block and the plane. An unknown constant force $\vec{F}$ is applied horizontally on the block as shown. The applied force is insufficient to support the mass. As a result the mass accelerates with magnitude a <u>down</u> the incline.

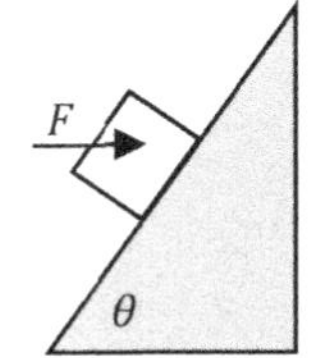

a) Draw an FBD and list the force equations for the block.
b) Determine the magnitude of the applied force.
c) Determine the normal force (magnitude) acting on the block.

5.11 Suppose you have several FBDs shown below. Assume all forces have identical magnitude F and the mass of the each object is m. Determine the components of the acceleration vector (a_x and a_y) for each object. To be clear, all force vectors lie in the xy-plane.

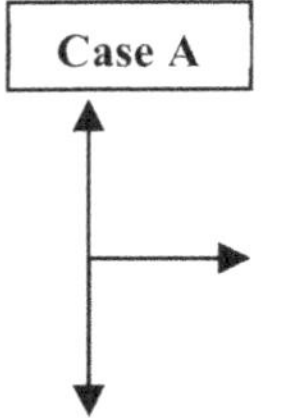

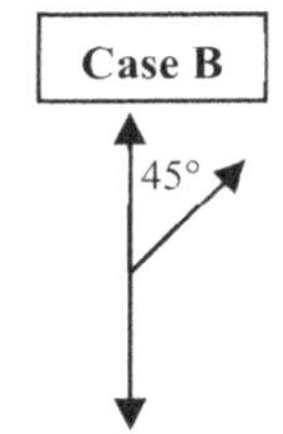

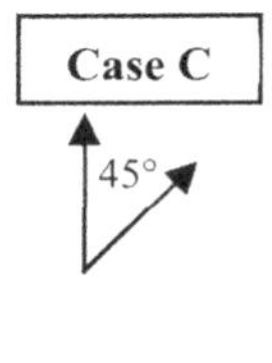

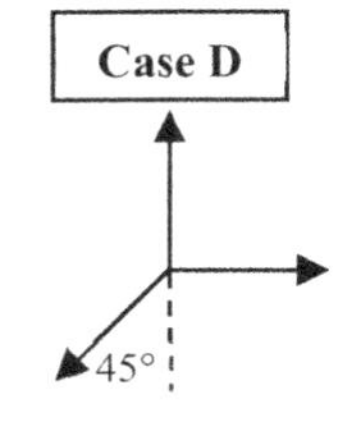

5.12 Two 1.0 kg masses are suspended on strings as shown in the figure. A scale connects the two strings above the table. This time someone pushes down on the left mass to set the system in motion. Assume $g = 10\frac{\text{m}}{\text{s}^2}$ for this problem.

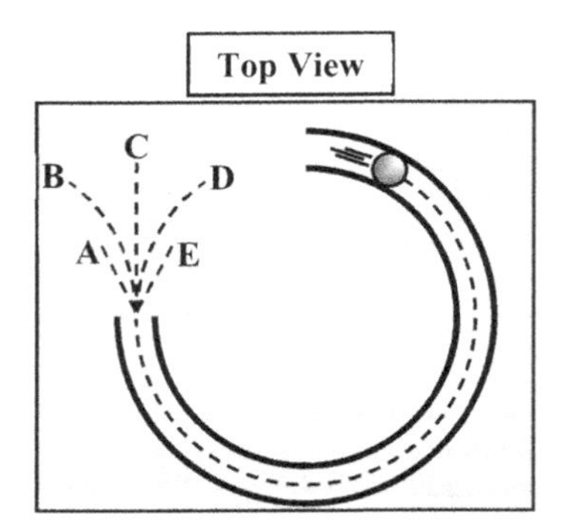

a) <u>After the push</u>, what will the scale read?

$T < 10$ N $\qquad$ $T = 10$ N $\qquad$ $T > 10$ N

b) <u>During the push</u>, what will the scale read?

$T < 10$ N $\qquad$ $T = 10$ N $\qquad$ $T > 10$ N

5.13 A marble is inside a tube that rests on a table. A top view of the tube is shown at right. The tube is bent so it forms ¾ of a circle. The marble is given a hard flick and rolls through the tube and comes out the other side. To be clear, gravity points into the page and the normal force points out of the page.
Which of the paths best approximates the trajectory of the marble upon exiting the tube? Support your answer using one of Newton's laws.

Top View

B C D A E

5.13½ Two forces act on a 2.50 kg object in deep space. Being in deep space, we needn't worry about any gravitational forces acting on the object. The object accelerates at a rate of $3.00\,\frac{\text{m}}{\text{s}^2}$ to the left. The first force has magnitude 12.50 N directed upwards. Determine the 2nd force (magnitude and direction) acting on the object. Include a sketch.

5.13 ¾ A spacecraft of mass m in deep space fires several thrusters at once. Each thruster exerts the same magnitude of force F. An FBD of the forces acting on the ship is shown below. The space ship starts from rest and accelerates at some *unknown angle* ϕ in the first quadrant. To be clear, because the spaceship is in deep space, we need not worry about the gravitational force mg.

a) Use the FBD shown at right to write down correct force equations in terms of the symbols F, θ, m, a_x & a_y.

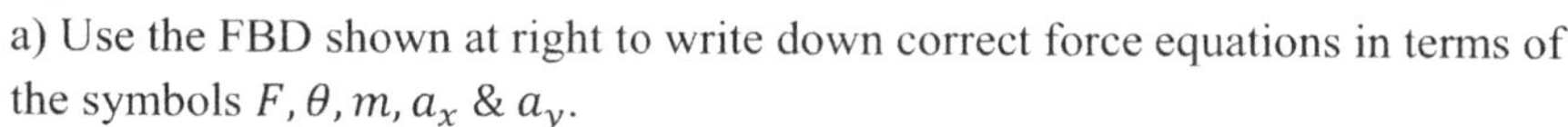

b) Determine the angle of acceleration ϕ. You should find the symbols F, a & m drop out. Furthermore, we know $\theta = 30.0°$. You should be able to get a number (with three sig figs) for the angle!!!

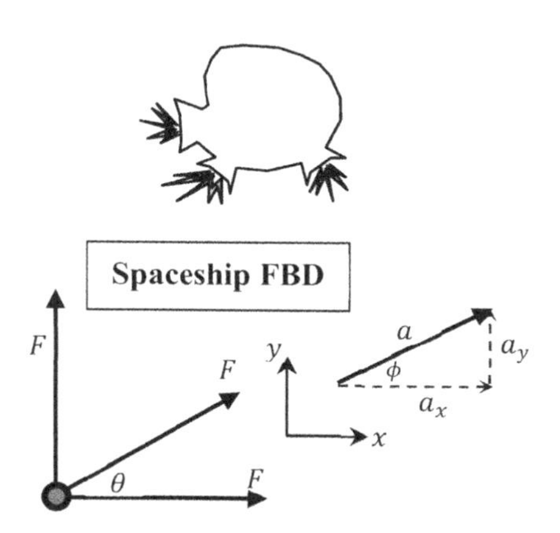

5.13 $\frac{7}{8}$ Three forces act on an object. The first force, caused by gravity, is $\vec{F}_1 = -19.6\text{ N}\,\hat{j}$. The second force has magnitude $F_2 = 20.0$ N pointing 30.0° *above the negative x-axis*. The third force is unknown. The object has acceleration $\vec{a} = (2.59\hat{\imath} - 8.75\hat{\jmath})\,\frac{\text{m}}{\text{s}^2}$. The object has mass $m = 2.00$ kg.

a) Determine the magnitude and direction of the third force.

b) Sketch and label all three forces acting on the object. Your sketch should be *roughly* to scale. For example, try to draw $\vec{F}_2$ *about* twice as long as $\vec{F}_1$ and ensure the 30.0° angle looks more like 30° than 45°. If it helps, use the grid below and assume each division represents 5 N.

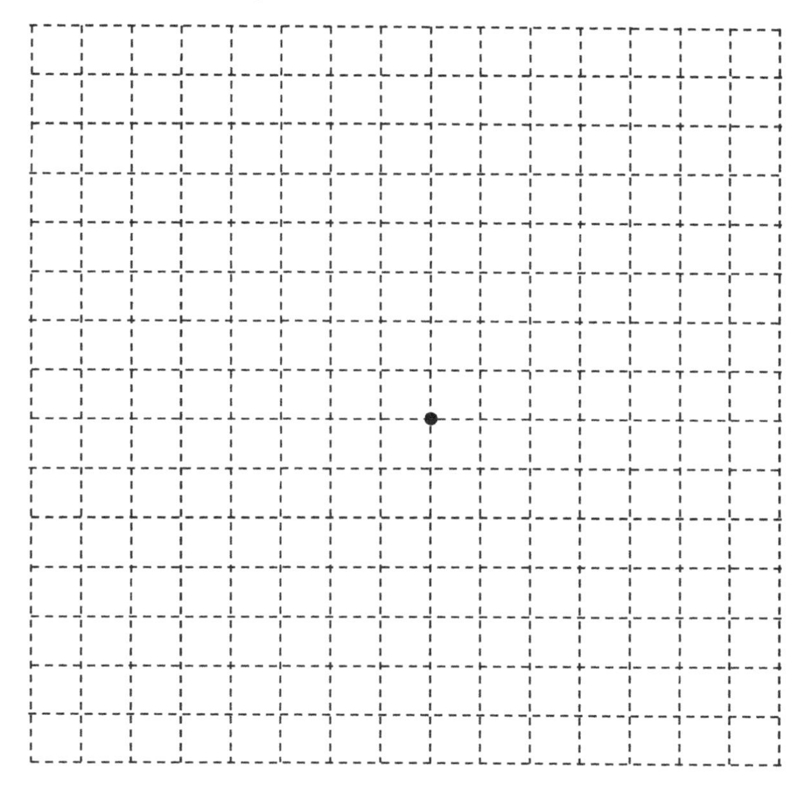

Newton's 3rd Law

- Force object 1 exerts on object 2 is equal in magnitude but opposite in direction to force 2 exerts on 1.
- $\vec{F}_{1on2} = -\vec{F}_{2on1}$ but also $F_{1on2} = F_{2on1}$. **Note the subtle difference!**
- If I push on the wall, the wall pushes back on me the same amount the opposite way.
- If a floor exerts friction to the left on a mop, the mop exerts friction to the right on the floor.
- Action-reaction forces are always the same type (both magnetic, both frictional, both gravitational, etc)

Action-Reaction Pairs

Oftentimes on tests I ask people to write down action-reaction pairs. What do I mean by this? Here is an example.

Suppose a block of cheese is dragged along the floor by a string. One action-reaction pair that relates to this problem is as follows:

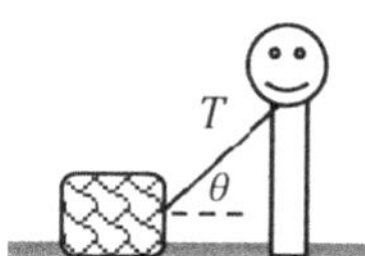

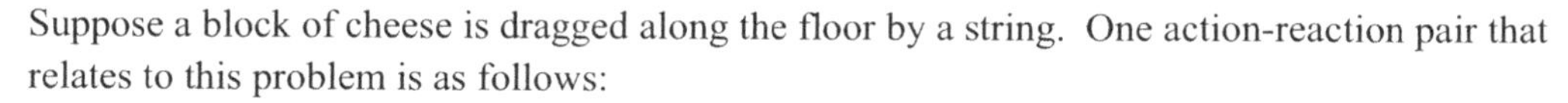

Action	Reaction
Zombie pulls up and to the right on the string using tension force.	String pulls down and to the left on Zombie using tension force.

Notice the following:

- Objects exert forces on other objects
- Objects do not exert forces on forces
- Action acts on object 1 while reaction acts on object 2.
- Action and reaction always act on different objects!
- Action and reaction are always point opposite directions.
- Action and reaction are always the same type of force.
- It doesn't matter if the objects are accelerating, at constant speed, or at rest.
- To get the reaction from an action, do the following:
  - Write down a sentence for the action which includes object 1, object 2, the direction of the force and the type of force.
  - Switch the objects in your sentence.
  - Switch the direction of the force in your sentence.
  - Leave the type of force the same.
    - Example: A exerts frictional force on B to the *right*. Therefore B exerts frictional force on A to the *left*.
    - Example: Earth exerts *downwards* gravitational force on zombie. Zombie exerts *upwards* gravitational force on earth.
- **Most common misconception:** normal force and weight are NEVER an action-reaction pair.
- Normal force and weight are NEVER action reaction pair.
- **Did I mention that normal force and weight are NEVER an action-reaction pair. Yes, I'm trying to shout at you. If I could slap you in the face with fish while I say this I would do it...anything to break this painful misconception. No fish face-slapping allowed.**

5.14 The sun exerts a gravitational pull on the earth directed towards the center of the sun. What is the other half of the action-reaction pair? Answer by stating which object exerts the force, the type of force (normal, frictional, electrical, etc), the direction of the force, and which object feels the force.

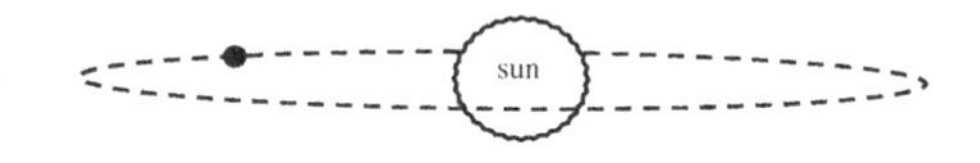

5.15 A block is at rest on the earth's surface. A normal force exerted by the earth acts upwardly on the block. A gravitational force exerted by the earth acts downwardly on the block. **What is the other half of the action-reaction pair associated with the gravitational force of the earth pulling the block down?** Answer by stating which object exerts the force, the type of force, the direction of the force, and which object feels the force.

5.16 A zombie exerts a horizontal force with magnitude F on a block as shown in the figure. The block accelerates parallel to the frictionless ramp with magnitude a. Describe an action-reaction pair associated with each of the three forces acting on the block. To be clear, you should be describing three pairs of forces for a total of six forces.

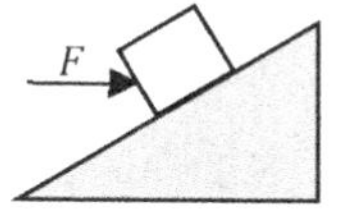

5.17 Imagine being in a car that accelerates from rest at an intersection.

a) What part of the car pushes you forward? Describe an action-reaction pair associated with this force.
b) What force pushes the car forwards? Describe an action reaction pair associated with this force.

5.18 Imagine being in a car that is taking a sharp turn. The turn is done in such a manner that your body presses against the door of the vehicle. Describe an action reaction pair associated with this force. Be sure to state which direction the door pushes on your body: towards the center of the circular motion or away from the center?

5.19 When you walk forwards, what force is pushing you forwards? Describe the action-reaction pair associated with this force.

5.20 Consider this process as a three stage event. First the person is in flight (stage 1). Then the person makes contact with the ground (stage 2). Then the person comes to rest (stage 3).

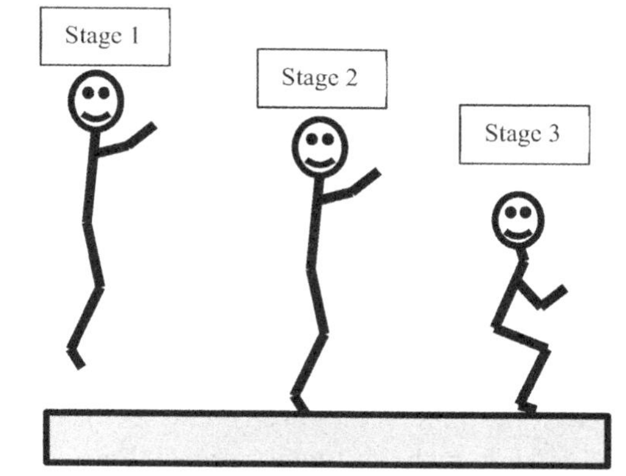

Note: in real life the normal force varies between stages 2 & 3. For the purposes of our problem, we can assume the normal force is essentially equal to the average value (n_{avg}) and assume it is approximately constant in magnitude.

a) Between stages 2 & 3, two forces act on the person: normal force and weight. Which force *magnitude* is larger on the person? Circle the best answer.
b) Between stages 2 & 3, the *earth* exerts a normal force on the *person* ($\vec{n}_{EonP}$). Simultaneously, the *person* exerts a normal force on the *earth* ($\vec{n}_{PonE}$). Which is *magnitude* is larger between stages 2 & 3? Circle the best answer.
c) We often discuss action-reaction pairs as described in lecture and the notes. Between stages 2 & 3, consider the weight of the person as the "action" force. What is the "reaction" force? Answer by stating 1) the object exerting the force, 2) the type of force (frictional, normal, gravitational, etc), 3) the direction of the force, and 4) the object experiencing the force. In other words, explain the reaction force by filling in the blanks in the following sentence if you expect to receive credit.

The ______________ exerts a __________________ force directed ______________ on the______________.

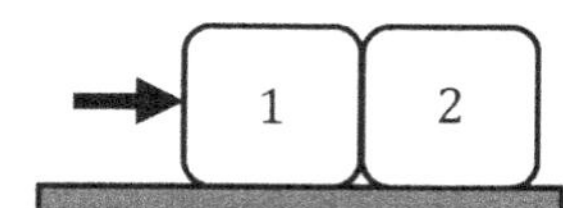

5.20½ Blocks 1 & 2 have masses $m_1 = 1.00$ kg & $m_2 = 2.00$ kg respectively. A zombie pushes horizontally on block 1 across a level surface. Block 1 exerts a normal force on block 2 with *magnitude* n_{1on2}. Block 2 exerts a normal force on block 1 with *magnitude* n_{2on1}.
a) If the blocks move to the right at constant speed, which normal force has larger magnitude?
b) If the blocks move to the right and *speed up* (assume $a_x > 0$), which normal force has larger magnitude?

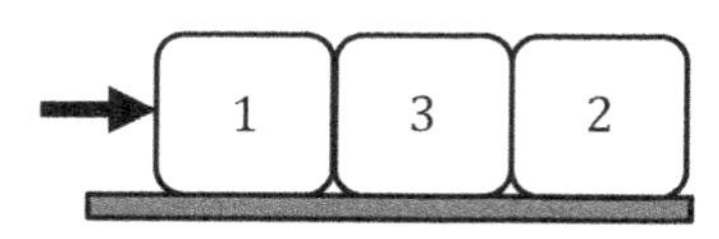

5.20¾ Now consider what happens when a third block of mass $m_3 = 3.00$ kg is placed between the same blocks 1 & 2. In this scenario, block 1 exerts a normal force *on block 3* with magnitude n_{1on3}. Block 2 exerts a normal force *on block* 3 with magnitude n_{2on3}. You may assume friction is negligible between block 3 and the floor.
a) If the blocks move to the right at constant speed, which normal force has larger magnitude?
b) If the blocks move to the right and *speed up* (assume $a_x > 0$), which normal force has larger magnitude?

5.21 A fan is attached to a cart that is free to roll on a track. Friction is negligible because the cart is on little wheels. The wheels are low mass and have low friction axles. As a result, this situation is almost perfectly modeled as a block that is free to slide with negligible friction. Initially the fan cart is at rest. The fan is activated. Explain why the cart moves by discussing an action-reaction pair involving the force that the fan exerts on the air.

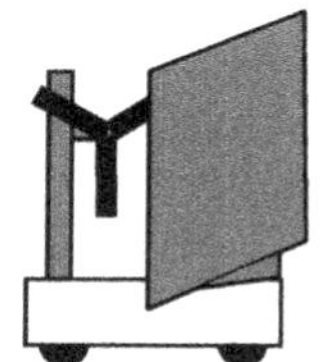

5.22 This time a goofy physics instructor puts a solid metal sail directly in front of the fan. The sail is essentially a large flat metal square.

a) Will the cart move or not?
b) Explain why using action-reaction pairs. This time discuss the action-reaction pair for the fan and the air AND the action-reaction pair for the sail and the air.

5.23 Finally, an even goofier physics instructor places a funky tube on the cart instead of a flat metal square. The tube is designed so the air from the fan enters the tube, reverses direction completely, then travels above the fan and flows out past it.

a) Will the cart move or not?
b) Which force should be larger: the one exerted by the air on the fan or the one exerted by the air on the tube? Ignore drag along the walls of the tube.

Get some light, flexible ducting from a hardware store and try it.

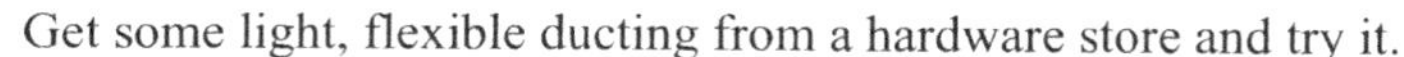

Is it possible to have propulsion in space without air to push against?
What if an astronaut opened the hatch on the space ship and starting throwing out supplies?
Suppose the astronaut pushes the supplies to the *left*.
By N3L, the suitcases push on the astronaut (and the spaceship) to the *right* (even in the vacuum of space)!
Similar logic holds true if a rocket thruster ejects exploded fuel to the *left*...

5.24 Challenge: Requires separation of variables: Is the fan cart moving with constant *power* or constant *force*?

a) Assume cart mass m & constant applied force with magnitude F. Determine an equation for the position as a function of time in terms of F and m. Assume the cart starts from rest with zero initial position.
b) While we will learn about power much later, it is sufficient for us to state that power is given by the equation $\mathcal{P} = \frac{\Delta Energy}{\Delta t} = \frac{\frac{1}{2}mv_f^2 - \frac{1}{2}mv_i^2}{\Delta t}$. If initial time and velocity are zero this becomes $\mathcal{P} = \frac{mv^2}{2t}$.
   From this we see the velocity as a function of time is $v = \sqrt{\frac{2\mathcal{P}t}{m}}$.
   Use this equation to determine the position as a function of time. Assume initial position is zero.
c) Think: what test could we do on a real fan cart to determine if power is constant or acceleration is constant?
   Note: this problem is revisited in chapter 7.

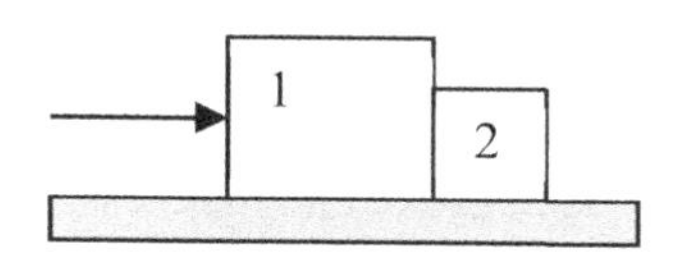

Systems and Internal/External Forces: Consider the two blocks pushed to the right in the top figure. They have the same acceleration and move together. We can be treat the two blocks as a single *system*. The normal force exerted by 1 on 2 is *internal* to the system. All other forces acting on either of the blocks are *external*. Notice the internal force tends to increase the speed of 2 while simultaneously opposing the motion of 1. Internal forces always come in action-reaction pairs and will always act in this way.

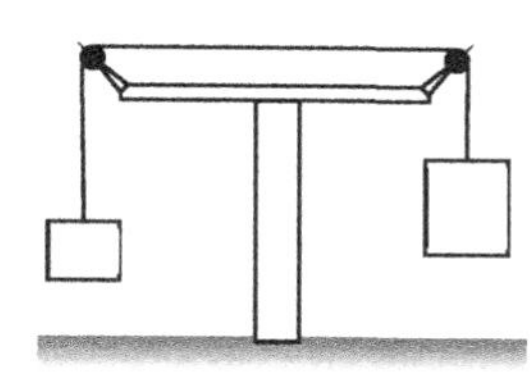

Consider the middle figure at right. The two blocks are connected by a string running over some pulleys. For now, assume the pulleys have negligible axle friction and are very light. Essentially, the pulleys do not impede the motion and serve only to bend the string. The two blocks accelerate in opposite directions but the magnitude of those accelerations is the same. In this sense, we can think of the two objects as both rotating in unison in a clockwise fashion. Considering the two blocks and the string a system, tension is now an *internal* force. Notice, again, the internal force tends to increase the speed of one block while simultaneously opposing the motion of the other block.

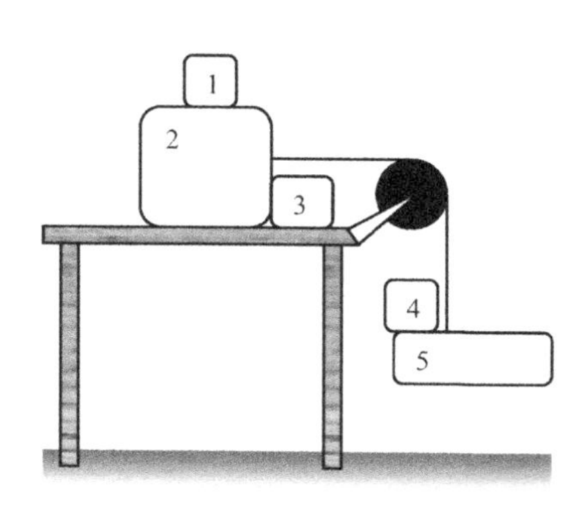

5.25 Consider the figure at right. Here we see a pulley system with stacked blocks. **You are told that friction prevents block 1 from sliding relative to block 2.** Now there are several interesting choices of systems. Depending on the problem, a wise choice of system can reduce algebra.

a) What forces are internal to the 4-5 system?
b) What forces are internal to the 1-2-3 system?
c) Define a system for which tension in the string is an internal force.
d) Suppose the friction was insufficient to keep block 1 on top of block 2. Freddy says it is still ok to treat 1-2-3 as a system because the friction between 1 and 2 will still be an action-reaction pair. Is Freddy correct? Explain.

Why care? Internal forces are ignored in a system FBD making the force equation easier to work with.

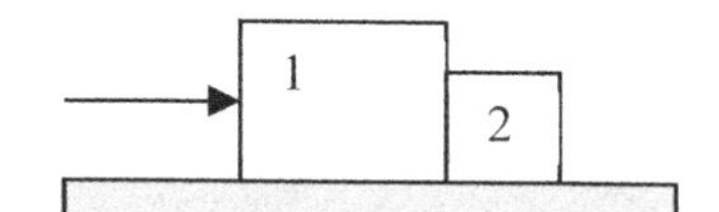

5.26 On a flat floor two blocks are placed side by side. Block 1 is pushed horizontally by force F and accelerates to the right. Assume all surfaces have negligible friction.

a) Draw an FBD for 1 and list the force equations.
b) Draw an FBD for 2 and list the force equations.
c) Draw an FBD for both blocks together (called the 1-2 system) and list the force equations.
d) Which FBD is best to determine the acceleration? Solve for a.
e) Which FBD(s) is best to find the normal force between the blocks? Solve for n_{12}.
f) Which force is internal and was left out of the 1-2 system FBD?
g) What is the action-reaction pair associated with the internal force?

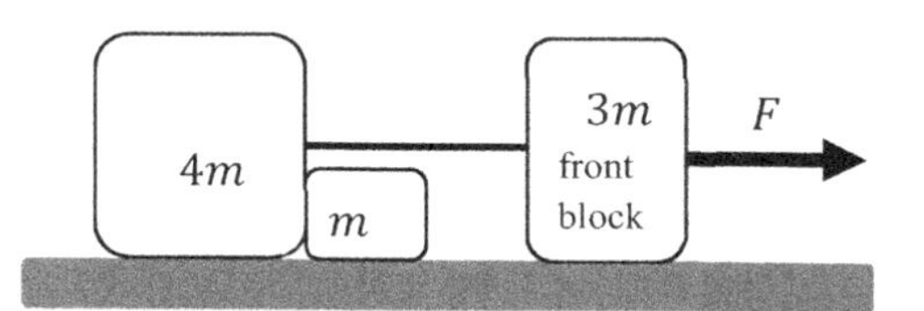

5.27 Three blocks are initially at rest on a horizontal surface. **All blocks experience negligible friction.** The mass of each block is labeled in the figure. You may assume these masses are known. A force of magnitude F is applied horizontally to the front block.

a) Draw FBDs and write *horizontal* force equations **for each block**.
b) Determine the acceleration magnitude of the front block. Answer in terms of F & m. Simplify any fractions or use decimal numbers with 3 sig figs.
c) Determine the magnitude of tension in the string between the two tall blocks. Answer either as a simplified fraction times F or with a decimal number with three sig figs times F.
d) Determine the magnitude of the normal force between the two trailing blocks. Answer either as a simplified fraction times F or with a decimal number with three sig figs times F.
e) Which forces are *internal* to a system which includes all three masses?

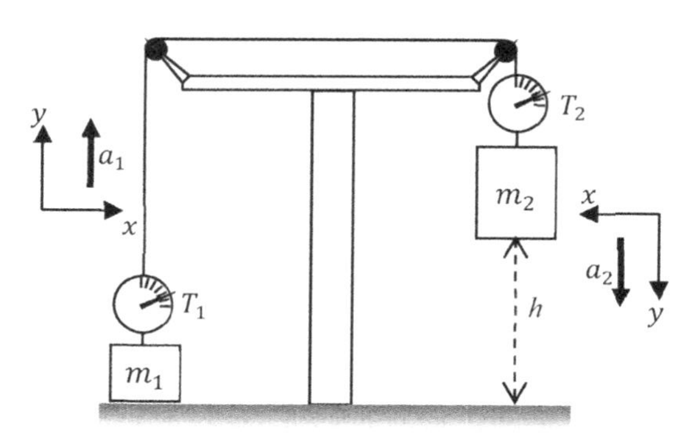

5.28 An Atwood's machine is shown at right. The pulleys are massless and have negligible friction at the axle. The string is inextensible and massless. Mass m_2 is larger than m_1. The system is released from rest when m_2 is distance h above the ground. The left scale reads the tension T_1 in the string above m_1 (similar for T_2). **Assume the masses of the scales are negligible.**

a) I'm assuming $a_1 = a_2$ for this problem. What key words in the problem statement make this assumption valid?
b) I'm assuming that the acceleration is constant for each mass. What key words in the statement make this assumption valid?
c) We will often assume $T_1 = T_2 = T$. What key words in the statement make this assumption valid?
d) Assume $a_1 = a_2 = a$ and $T_1 = T_2 = T$. Draw separate FBDs for m_1 and m_2 using the coordinates shown. Determine force equations for each block separately.
e) Determine a in terms of m_1, m_2, and g. Write your answer as $a = g\frac{?}{?}$. This makes it easy to check the units.
f) Determine an expression for T in terms of m_1, m_2, and g. Check the units.
g) Do your expressions for a and T make sense when $m_2 = m_1$? Explain.
h) Do your expressions for a and T make sense when $m_2 \gg m_1$? Explain.
i) Do your expressions still make sense if $m_2 < m_1$? Explain.
j) Considering motion relative to the ground, does $\vec{a}_1 = \vec{a}_2$? Is it better to say $\vec{a}_1 = -\vec{a}_2$? Explain.
k) If both blocks and the connecting string are viewed as a system, which force is an "internal force"?
l) Suppose we measure the time to fall t and measure the height h. We could get an experimental value for a to compare to the Newton's 2nd law theory. Use kinematics to determine an expression for a_{exp} in terms of h and t. Note: the reading on the scales gives a direct value for T_{exp}.

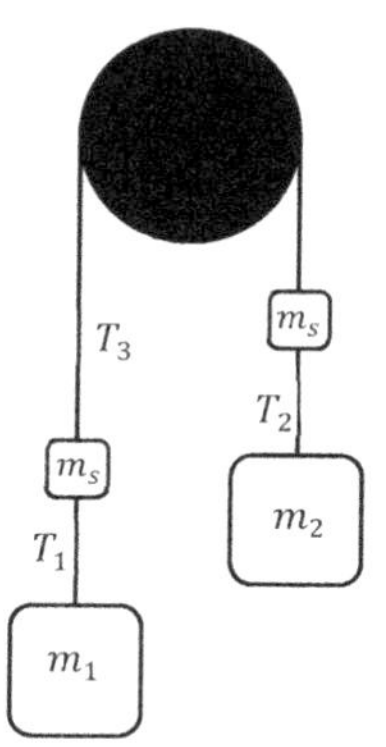

5.29 We could include the scales in the FBDs. Consider the figure at right. Each string gets its own tension since the scale mass (m_s) must now be accounted for in the FBDs.

a) Determine the acceleration magnitude and each tension of this system.
b) Beyond checking the units, it is appropriate to verify the result of this problem matches up to the result of the previous problem when $m_s = 0$. A solution that does not match the simpler problem cannot be correct. A solution that does match is more trustworthy.

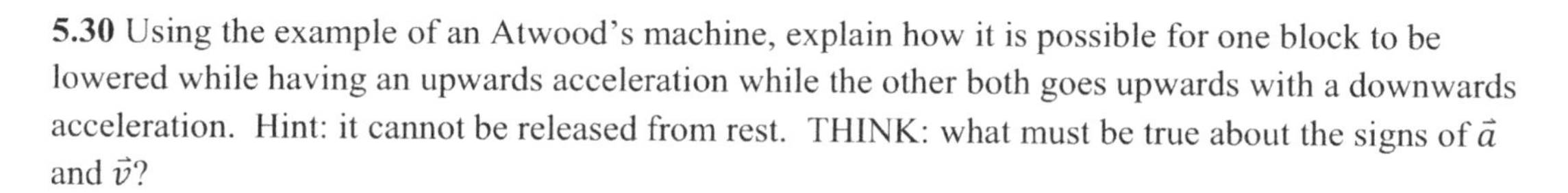

5.30 Using the example of an Atwood's machine, explain how it is possible for one block to be lowered while having an upwards acceleration while the other both goes upwards with a downwards acceleration. Hint: it cannot be released from rest. THINK: what must be true about the signs of $\vec{a}$ and $\vec{v}$?

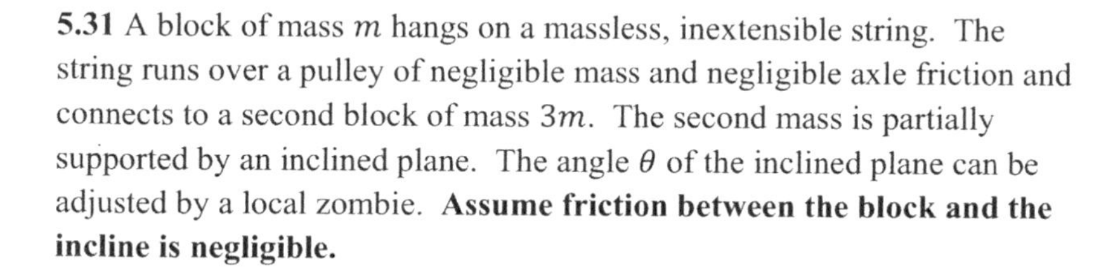

5.31 A block of mass m hangs on a massless, inextensible string. The string runs over a pulley of negligible mass and negligible axle friction and connects to a second block of mass $3m$. The second mass is partially supported by an inclined plane. The angle θ of the inclined plane can be adjusted by a local zombie. **Assume friction between the block and the incline is negligible.**

At what angle will the two blocks be in equilibrium? Hint: you should be able to get a number for the angle with 3 sig figs.

5.32 Four blocks are connected by three strings as shown in the figure below. A zombie pulls the right most block at some angle shown in the figure. Assume the masses and angles are all known. Assume friction is negligible between all surfaces. Assume the rightmost mass actually travels to the right.

a) Determine the acceleration (magnitude) in terms of the masses, the magnitude of zombie force F, the magnitude for freefall acceleration g, and the angles.
b) Assuming all masses are now identical, determine the minimum zombie force magnitude required to make the system of blocks motionless. Answer in terms of m, g, θ, and ϕ.
c) Think about the force equations in the vertical direction for block 4. This force equation relates the tension and angle to the weight of block 4. I assumed block four remained on the floor for the entire problem. Notice this constrains either the force magnitude or angle for the applied zombie force. Not every zombie force is realistically modelled by the work for the above parts.

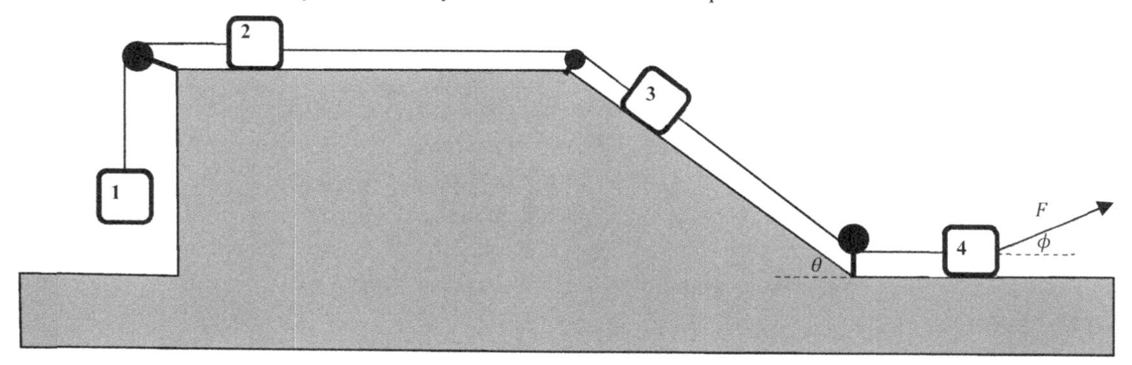

5.33 A girl of mass m_1 sits in a chair of mass m_2 supported by rope that runs over a pulley to her hand. The inextensible rope may be considered massless compared to the girl and chair. By pulling down on the rope with her hand she accelerates upwards with magnitude $a = \frac{g}{10}$.

a) Determine the tension (magnitude) in the rope.
b) Determine the normal force (magnitude) exerted by the seat on the girl.
c) Determine the net force exerted on the ceiling. You may assume the mass of the pulley is negligible.
d) Can you think of any special cases we could use to check our answers? Do your answers seem reasonable? Are the answers what you expect?

5.34 In case 1 a zombie exerts a force <u>parallel to the incline</u> with magnitude F on a block as shown in the figure. In case 2 zombie exerts a <u>horizontal</u> force with the same magnitude on an identical block. The mass of each block is m.

a) Determine the ratio of the acceleration (magnitude) in case 1 to case 2. Think: which should be larger? Should the ratio be greater than 1, less than 1, or equal to 1?
b) Determine the ratio of the normal force (magnitude) in case 1 to case 2. Think: which should be larger? Should the ratio be greater than 1, less than 1, or equal to 1?
c) How should your results change as the angle goes to zero?
d) Which case makes no sense if the angle goes to 90°? Explain why?

CASE 1

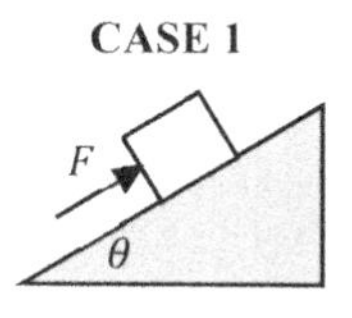

CASE 2

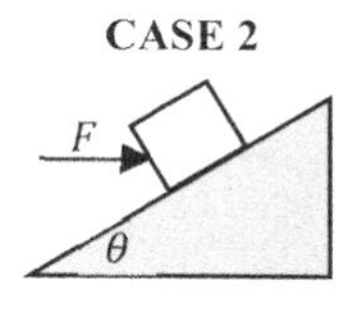

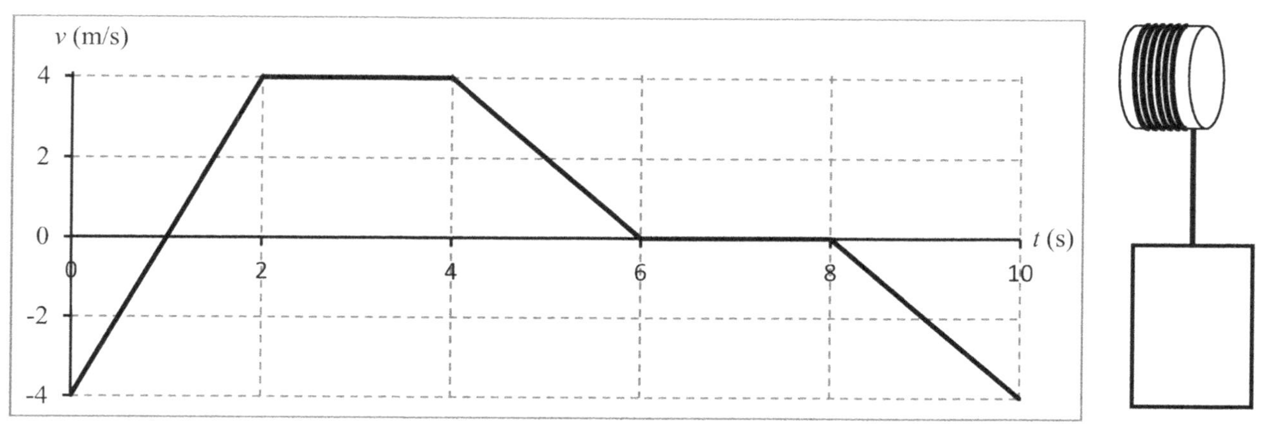

5.35 A clever student uses an app on their phone to generate a plot of velocity versus time for an empty elevator (plot shown directly above). You may think of the elevator as a box of mass m attached to a cable. The other end of the cable wraps around a winch attached to the ceiling (see figure above right).

a) During what time interval(s), or at what time(s), is the elevator moving down and speeding up?

b) During what time interval(s), or at what time(s), is the elevator at rest?

c) During what time interval(s), or at what time(s), is the tension in the cable <u>less than</u> mg?

d) During what time interval(s), or at what time(s), is the tension in the cable <u>equal to</u> mg?

e) Suppose the elevator is initially at the third floor which is a height of 20.0 m above the ground. Assume the ground is considered $y = 0$. Write an equation for vertical position as a function of time valid for the first 2.0 seconds of motion.

Student observation: this elevator is doing some pretty crazy moves…perhaps it is a ride at an amusement park?

5.36 Consider the figure at right. The zombie pulls on a block of cheese. Assume there *is* a friction force between the cheese and the floor. We do not know if the cheese is accelerating, at constant speed, or at rest. There are four forces acting on the block of cheese. Write down an action-reaction pair for each force.

5.37 Describe a specific physical scenario wherein an object is being lowered yet has an upwards acceleration. Is this even possible? Explain why it is not possible <u>**OR**</u> give a specific example of how it is possible.

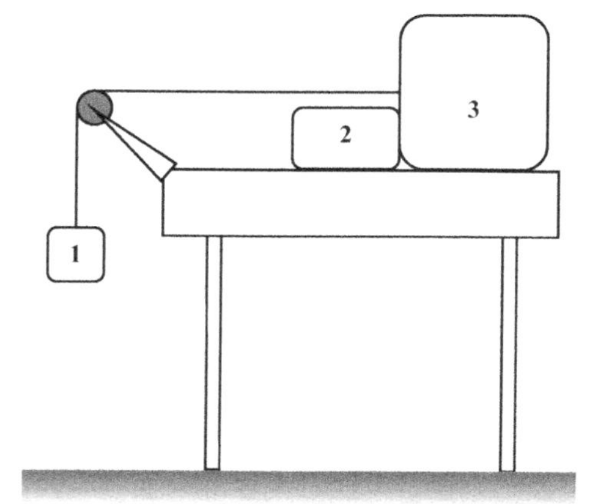

5.38 Three blocks are connected as shown in the figure. Assume the string is massless and inextensible. Assume the pulley is massless and frictionless. Blocks 2 & 3 are in contact and are supported by a horizontal table. For this problem friction is negligible. Assume blocks 1, 2, & 3 have masses m, $2m$, & $3m$ respectively.

a) Determine the tension (magnitude) in the string. Write your answer as simplified fraction times mg.

b) Determine the normal force (magnitude) between blocks 2 & 3. Write your answer as simplified fraction times mg.

c) While accelerating, which does block 2 exert more normal force on block 3 or is it the other way around?

5.39 Numerical Consider the Atwood's machine shown at right. This scenario is identical to problem 5.21. In that problem we found

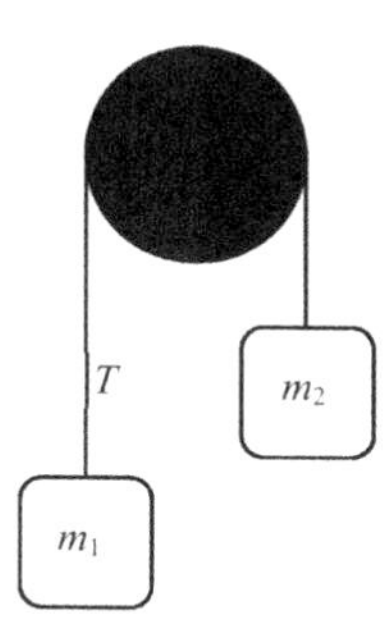

$$a = g\frac{m_2 - m_1}{m_2 + m_1} \qquad\qquad T = \frac{2m_1 m_2 g}{m_2 + m_1}$$

Let us now assume that $m_1 = 1$ kg and $g = 10\,\frac{\text{m}}{\text{s}^2}$. The other mass is free to vary and will take on a range of different values between 0 and 20 kg.

a) Fill in the table for the three special values of m_2 shown below.

m_2 (kg)	$a\left(\frac{m}{s^2}\right)$	T (N)
0		
1		
20		

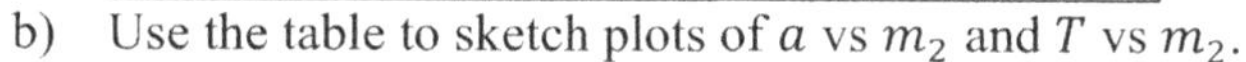

b) Use the table to sketch plots of a vs m_2 and T vs m_2.

c) Use Excel, Matlab, or some other program to create a table of values and make plots. Compare them to your sketch. I believe both techniques (sketching plots or using software to make plots) are useful.

d) **Challenge:** For fun, consider the quantity $\frac{T}{m_2 g}$. In some sense this represents which force, weight or tension, dominates the behavior of m_2. The larger the number, the more tension dominates and pulls m_2 upwards. What is the limit of $\frac{T}{m_2 g}$ as $m_2 \to 0$?

5.40 Numerical Consider m_2 as a hanging mass while m_1 is at rest on a frictionless incline. The system is released from rest.

a) Assuming up the plane is positive, show:

$$a = g\frac{m_2 - m_1 \sin\theta}{m_2 + m_1}$$

$$T = \frac{m_1 m_2 (1 + \sin\theta) g}{m_2 + m_1}$$

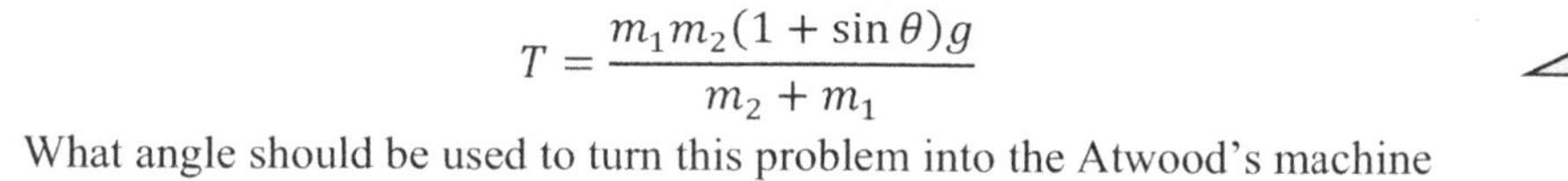

b) What angle should be used to turn this problem into the Atwood's machine from problem **5.28**? Verify the equations are correct in the Atwood's machine limit.

c) Now assume $m_1 = 2$ kg, $m_2 = 1$ kg and $g = 10\,\frac{\text{m}}{\text{s}^2}$. The angle is free to vary between 0 and 90°. What angle will put the system into equilibrium?

d) Sketch plots of a vs θ and T vs θ. If it helps, fill in the table below.

θ (°)	$a\left(\frac{m}{s^2}\right)$	T (N)
0		
30		
90		

e) Now use software to create the plots. Hint: Excel computes sine and cosine using radians.

f) **Challenge:** make a contour plot showing the acceleration as both m_2 and θ are varied.

Constrained motion with pulleys

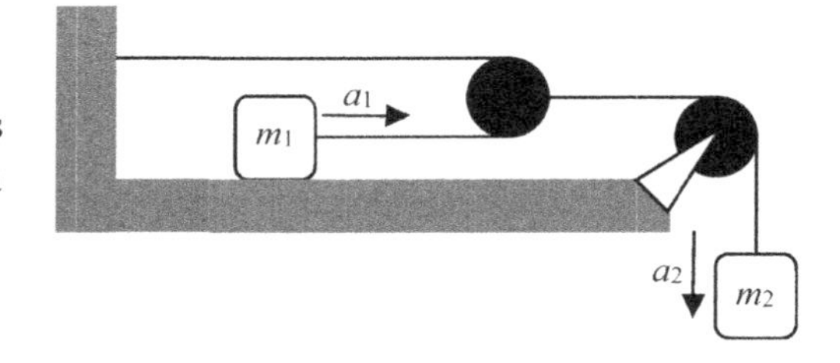

5.41 Challenge/Mechanical Engineers Two blocks are connected using the pulley system shown at right (figure not to scale). For all intents and purposes, the pulleys are massless with frictionless axles and the string is massless and inextensible. Block m_1 slides along a table experiencing negligible friction. The system is released from rest and m_2 falls towards the floor.

a) Suppose m_2 falls distance h. How far will m_1 move?

   h $\qquad$ $2h$ $\qquad$ $h/2$

b) Based on your previous answer, how should a_1 relate to a_2?

   $a_1 = a_2$ $\qquad$ $a_1 = 2a_2$ $\qquad$ $a_2 = 2a_1$

c) Draw an FBD for m_1, the *left* pulley, and m_2. Write down the force equations. Derive an expression for a_2 in terms of g, m_1, and m_2. Hint: at some point you should have the mass of the pulley (m_p) appear in one of your force equations. Since m_p is negligible compared to the other masses you may set $m_p = 0$. Determine equations for each acceleration and tension.

d) Check your work by considering the following special cases:

   i. $m_1 \to 0$
   ii. $m_1 \to \infty$
   iii. $m_2 \to 0$
   iv. $m_2 \to \infty$

If you are looking to do this problem as a demonstration in class consider the following scenario:

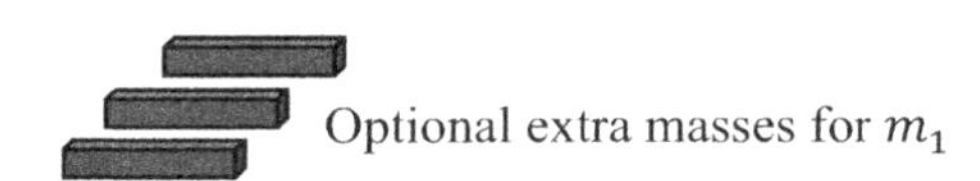

Mass m_1 is a cart with low-friction wheels with mass 500 g. Each rectangular black mass is approximately 500 g as well. In my experience using $m_2 = 250$ g worked fine for a cart with three black rectangular masses stacked on it. At the left end of the track a light, inextensible string is tied to a brick, extends over to a pulley, then comes back to connect to m_1. The brick will stay locked in place. A different string attaches to the mounting bracket of the pulley at one end and to m_2 at the other end. Both pulleys have negligible mass and friction. The system is released from rest. The an experimental value of the acceleration is given by $a_{2exp} = g\frac{2h}{t^2}$

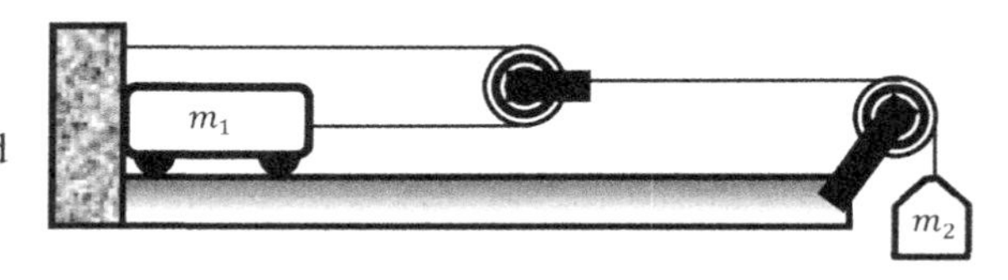

Error Analysis: The masses, h, and g are found to a precision of 1% of better. The timing typically is on the order of 2 seconds for a 1 meter fall. Even with efforts used to decrease the effects of human reaction time, one expects reaction time to cause errors on the order of 0.1 seconds. This means reaction time error is approximately 5%. Since time is squared in our equations, the error is doubled to about 10%. Overall, we expect our errors to be approximately 10%.

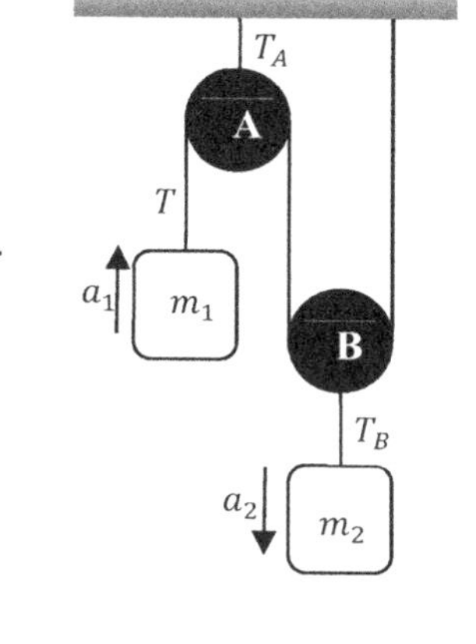

5.42 Challenge/Mechanical Engineers: Two blocks are connected using the pulley system shown at right (figure not to scale). The pulleys are low-mass with low-friction axles. The string used is light with negligible stretch. Block m_1 is connected to the ceiling by a string that wraps over pulley **A**, then around pulley **B**. A string with tension T_A connects the center of pulley **A** to the ceiling. A string with tension T_B connects m_2 to the center of pulley **B**. The system is released from rest and m_2 falls towards the floor.

a) Determine an equation relating the magnitudes of acceleration a_1 and a_2.
b) Draw free body diagrams and determine force equations for m_1, m_2, and each pulley. Note: mg and ma terms drop out since mass of the pulley is negligible!
c) Determine equations for each acceleration and tension as well as the total downward force from the system on the ceiling.
d) Check your work by considering the following special cases:
   i. $m_1 \to 0$
   ii. $m_1 \to \infty$
   iii. $m_2 \to 0$
   iv. $m_2 \to \infty$
   v. $m_2 = 2m_1$

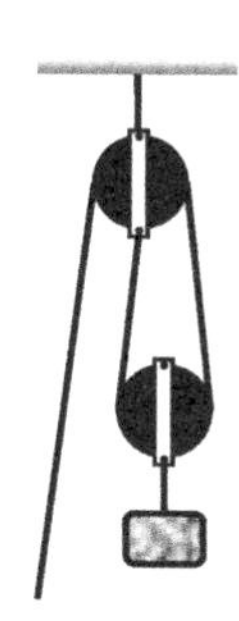

You might find it interesting to check out "block and tackle" or "Bosun's Chair" online. You will find there are many ways to use multiple pulleys to reduce the force required to lift a block. As we will learn later, decreasing the amount of force required does not come for free. In exchange for requiring less force to lift a block, a longer section of rope must be pulled. For example, in the system shown at right, the force required to lift the block is cut in half while the amount of rope that must be pulled is doubled. Here I am ignoring the slight angles of the ropes and assuming all rope sections are purely vertical. As long as the angle from the vertical is less than about 8°, the vertical string model introduces errors of less than 1%.

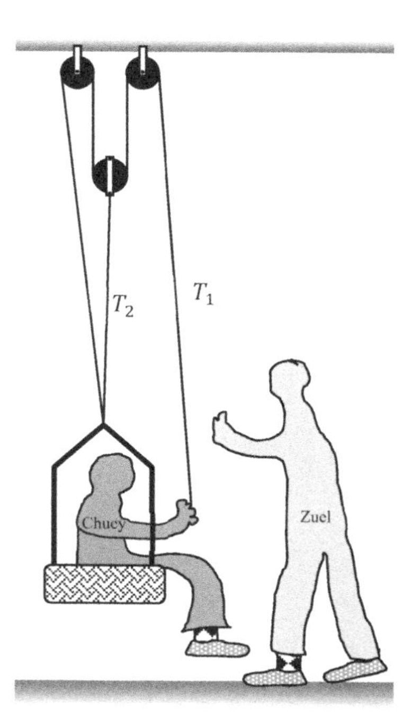

5.43 Part a good for all, the rest is Challenge for ME's: As a fun thing to think about, consider the pulley chair shown at right. Sometimes science museums will have chairs like this with different numbers of pulleys. The point is to help people feel difference as more pulleys are added.

Chuey and Zuel visit the local science museum. Chuey sits in the pulley chair and holds himself at rest suspended above the ground.

a) Make a list of assumptions used to simplify this problem. State not only the assumption but also why you are making it (how does it simplify the model). Don't overthink it, just do this as quickly as you can then compare to the solution.
b) What is the total force on the ceiling?
c) What is the tension in each cable?
d) If Zuel is holding the rope instead of Chuey, how do the above answers change? Increase, decrease, stay the same, or is more information needed?
e) How do the tensions change if you are told Chuey is being lowered back to the ground? Increase, decrease, stay the same, or is more information needed?
f) If *no one* holds the rope, at what rate will Chuey accelerate downwards?

More constraint problems?

Assume each pulley has negligible axle friction. Determine acceleration of each mass and tension in the cable. No solutions available, but fun to think about.

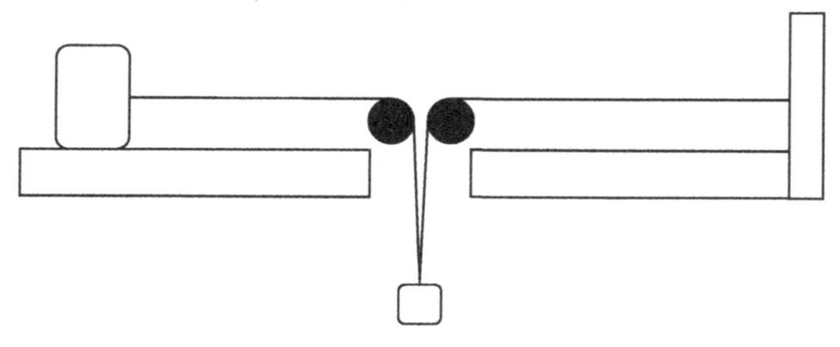

Figure out the tension in all three cables and the acceleration of all masses. No solutions available, but fun to think about.

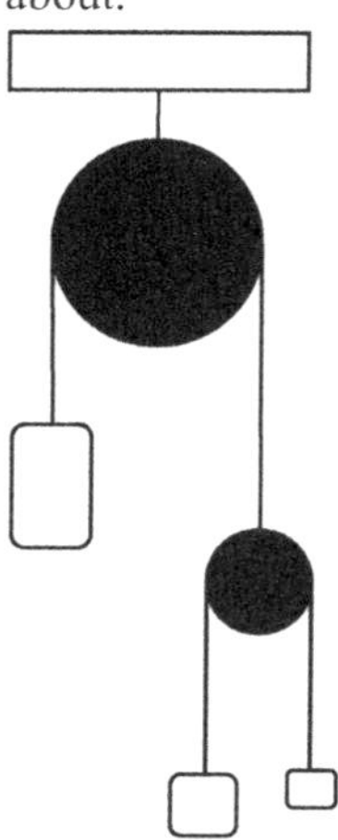

FRICTION

In real life friction can be quite complicated. For a great many cases, however, a simple model for friction does quite well. In this simple model we find that frictional forces do not depend on the size of the area in contact with the surface. This model works fine for things like blocks, books, boxes, etc on tables, floors, etc. The model is split into cases below.

Friction cases are distinguished by considering the relative motion of the surfaces involved.

1) For two surfaces **in relative motion** $f = \mu_k n$.
2) For two surfaces **not in relative motion**
   a) The maximum possible friction force occurs when the objects are on the verge of slipping. **On the verge of slipping**, assume $f = \mu_s n$. If a problem mentions "just about to slip", "largest angle without sliding", "minimum mass to prevent slipping", "minimum speed without sliding", or something similar you may assume $f = \mu_s n$.
   b) Some problems do *not* indicate things are on the verge of slipping. In these problems I like to say $f\ is\ what\ it\ is$. **Resist the temptation to use $f = \mu_s n$.** In these problems you often are asked to determine f. At the end of the problem you can verify $f < \mu_s n$ but that's about it.

There are a few points worth emphasizing:

- In general $n \neq mg$. Do the FBD and force equation to determine n.
- In general $n \neq mg\cos\theta$. Do the FBD and force equation to determine n.
- Friction usually depends on the normal force (*but not always*). Remember case 2b.
- Typically μ_s is larger than μ_k. In some problems we ignore this difference and assume $\mu_s = \mu_k = \mu$.
- **Friction is not μ.** On free body diagrams write f (the frictional force) not μ (the frictional coefficient).

Which way does friction point?
You might be wondering why I would bother to put this in here. Shouldn't this be obvious? Evidently not. To be clear, friction, in general does not point opposite the acceleration!

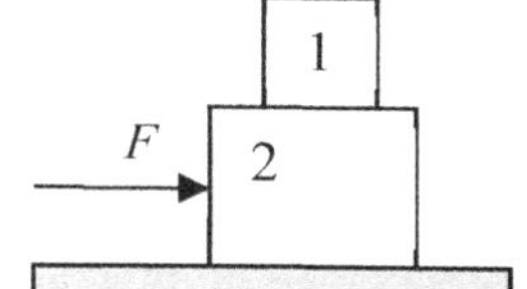

Consider the example at right. Block 2 is pushed along a table and accelerates to the right. Assume block 1 remains on top of block 2 (it doesn't slide relative to block 2). Note this chain of reasoning:

- Clearly block 1 must also accelerate to the right (1 not moving *relative to 2*).
- There is only one force acting horizontally *on block 1*: friction *from 2 on 1*.
- Therefore friction *from 2 on 1* points to the right (else it wouldn't accelerate to the right).
- Furthermore, unless we are told the object is on the verge of slipping, we expect $f < \mu_s n$.
- More specifically, it is inappropriate to use $f = \mu_s n$ for this scenario!
- Newton's 3rd law: friction *from 1 on 2* points opposite friction *from 2 on 1*(friction *from 1 on 2* points left).
- Since block 2 slides to the right relative to the floor, friction from the floor on block 2 is $f = \mu_k n$ and points to the right.

I like to think of it this way: first consider the relative motion (or tendency to move) between two objects. Let this direction guide your reasoning as to the direction friction points.

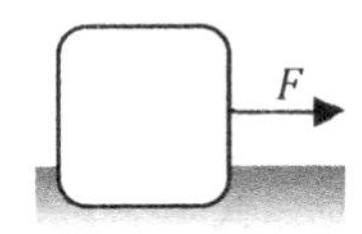

6.1 A block of mass $m = 1.00$ kg is at rest on a flat surface. The coefficients of friction are $\mu_s = 0.5$ and $\mu_k = 0.3$. A force with magnitude F is applied horizontally to the block. The force is gradually increased from 0 to 10.0 N. Even though I am sloppy with sig figs in the problem statement, put 3 sig figs on all answers.

a) At some point, the block will slip. Let us first determine the largest force we can apply without slipping. Draw and FBD, determine the force equations, and determine the largest force we can apply before the block begins to slide. Said another way, figure out the minimum magnitude F_{min} required to cause slipping.

b) Now assume the applied force is $F < F_{min}$. In particular, assume $F = 3.00$ N. Determine f and a. Consider the following:

   i. Does the FBD change?

   ii. Can you still use $f = \mu_s n$?

c) Now assume the applied force is $F > F_{min}$. In particular, assume $F = 4.91$ N. Determine f and a.

   i. Does the FBD change?

   ii. Can you still use $f = \mu_s n$?

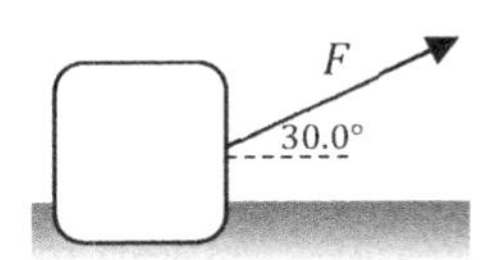

6.2 A block of mass $m = 1.00$ kg is at rest on a flat surface. The coefficients of friction are $\mu_s = 0.5$ and $\mu_k = 0.3$. A force with magnitude F is applied at an angle of 30.0° to the block. The force is gradually increased from 0 to 10.0 N. Even though I am sloppy with sig figs in the problem statement, put 3 sig figs on all answers.

a) Take a guess: will the block slide for larger or smaller values of F? Said another way, is F_{min} going to be bigger, smaller or the same as in the last problem?

b) At some point, the block will slip. Let us first determine the largest force we can apply without slipping. Draw and FBD, determine the force equations, and determine the largest force we can apply before the block begins to slide. Said another way, figure out the minimum magnitude F_{min} required to cause slipping.

c) Now assume the applied force is $F < F_{min}$. In particular, assume $F = 3.00$ N. Determine f and a. Consider the following:

   i. Should the number differ from the previous problem's part b?

d) Now assume the applied force is $F > F_{min}$. To compare to the previous problem, assume $F = 4.91$ N. Determine f and a.

   i. Should the number differ from the previous problem's part c?

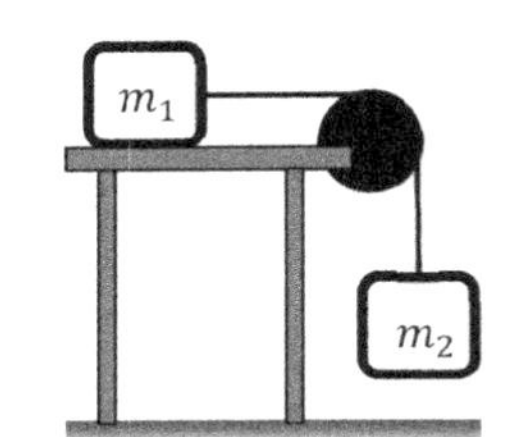

6.3 Block m_1 rests on a table. Coefficients of friction between m_1 & the table are μ_s & μ_k. A light, inextensible string connects m_1 to m_2 over a light pulley (negligible axle friction).

a) Determine the minimum mass m_{2min} required to cause slipping.

b) Determine acceleration magnitude when $m_2 = 2m_{2min}$.

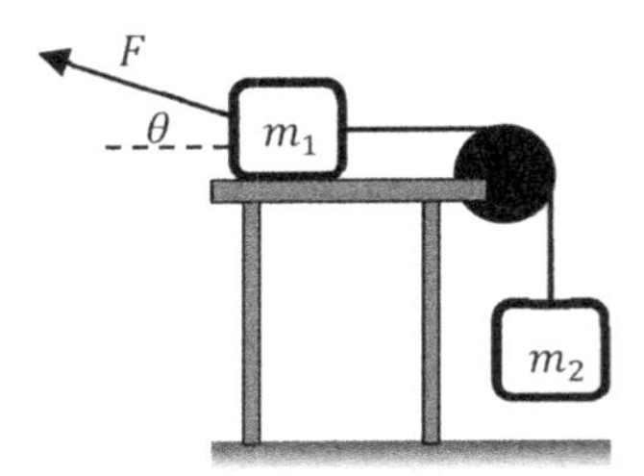

6.3½ Block m_1 rests on a table. Coefficients of friction between m_1 & the table are μ_s & μ_k. A light, inextensible string connects m_1 to m_2 over a light pulley (negligible axle friction). A force with variable magnitude F is applied at fixed angle θ.

a) Determine the minimum force magnitude (F_{min}) required to cause slipping.

b) Determine acceleration magnitude when $F = 2F_{min}$.

For the following situations determine the direction of friction. Also determine if it is appropriate to model friction as $f = \mu_k n$, $f = \mu_s n$, or $f < \mu_s n$. If more than one frictional/normal force appears in the equation, determine the direction and appropriate friction case for all frictional forces. Do not worry about FBDs and force equations now.

6.4 A block rests on a horizontal table. The frictional coefficients between the block and table are μ_s & μ_k.

6.5 A block slides to the right when it is pulled by a zombie. The frictional coefficients between the block and the floor are μ_s & μ_k.

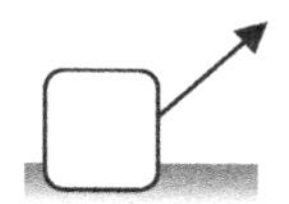

6.6 Determine the largest m_1 that can be used without m_2 sliding up the plane. The coefficients between all surfaces are μ_s & μ_k.

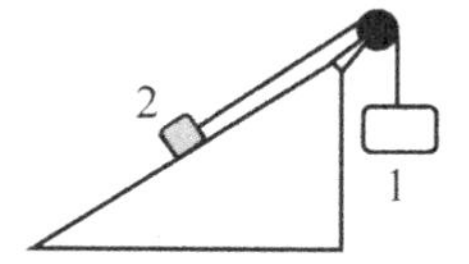

6.7 A box sits on a flat bed truck. The truck accelerates from rest with magnitude $a = \frac{g}{5}$. The coefficients between the block and truck are μ_s & μ_k. Determine the frictional force on the box.

6.8 Two boxes are stacked on top of each other as shown in the figure. The coefficients between all surfaces are μ_s & μ_k. A zombie pushes horizontally on the lower block. Determine the largest acceleration for which the stacked boxes will continue to move together.

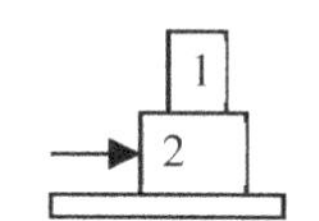

6.9 Two boxes are stacked on top of each other. The coefficients between all surfaces are μ_s & μ_k. A zombie pulls horizontally on the upper block causing the blocks to move together. Determine the acceleration.

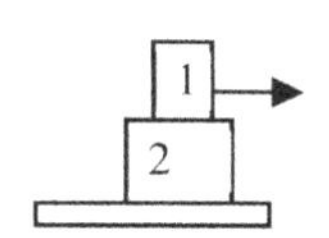

6.10 A zombie is on a carnival ride. The ride is essentially a giant spinning cylinder with rotation rate ω. The woman is pressing up against the wall and, as Newton's 3rd law tells us, the wall pushes her towards the axis of rotation. In addition, the wall supports her so that her feet do not touch the ground. Determine the net force exerted by the wall on the zombie (magnitude and direction).

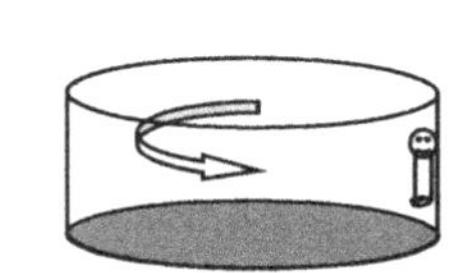

6.11 A small force is directed at angle θ on a block of mass m as shown in the figure. The magnitude of the force (F) is held constant while the angle θ is gradually increased. The block remains stationary the entire time.

a) As θ increases, how does the normal force change? Increase, decrease, or constant.
b) As θ increases, how does the frictional force change?
c) As θ increases, how does the maximum possible frictional force change?

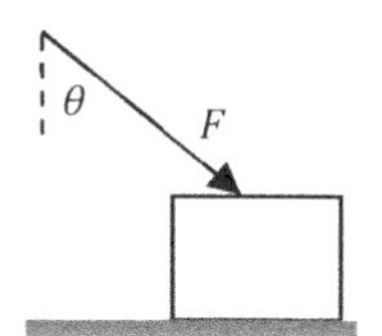

6.12 Now reconsider the problem of a box on a flat-bed truck that accelerates from rest. Suppose the truck accelerates so quickly that the box begins to slide off the back.

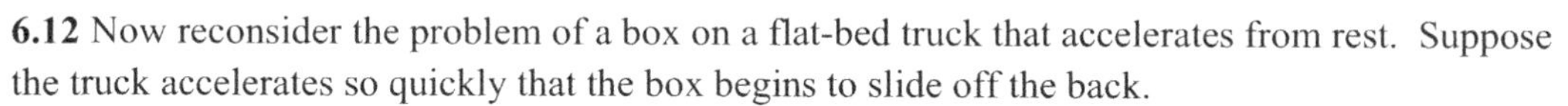

a) Compare the direction of friction (on the block) to block velocity relative to the truck?
b) Compare the direction of friction (on the block) to block velocity relative to the earth?

6.13 A block is held in place on a ramp by horizontal force with magnitude F. The coefficients between the block and the ramp are μ_s & μ_k.

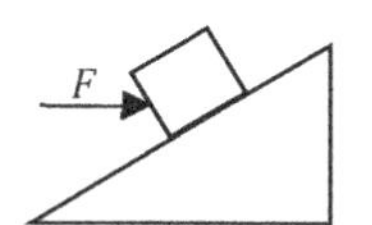

6.14 Assume a block of mass m is at rest on an incline. The angle is increased until the block is just about to slide. This special angle is called the critical angle (θ_c).

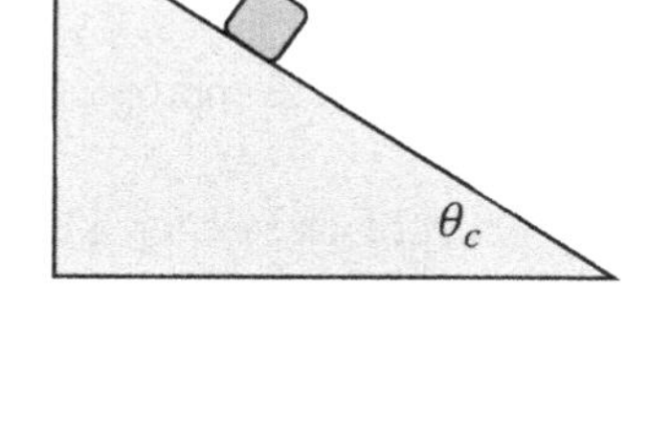

a) Which case for friction applies at θ_c?
b) Sketch the FBD and determine the force equations at θ_c.
c) Solve for the appropriate coefficient of friction in terms of θ_c.
d) Suppose a block with twice the mass is used. How is the critical angle affected?
e) For angles <u>above</u> the critical angle, does $f = \mu n$? If so, would one use μ_s or μ_k?
f) For angles <u>below</u> the critical angle, does $f = \mu n$? If so, would one use μ_s or μ_k?

6.15 A block of mass m is released from rest on an incline. The angle θ of the incline is greater than the critical angle. The block slides a distance L down the incline in time t.

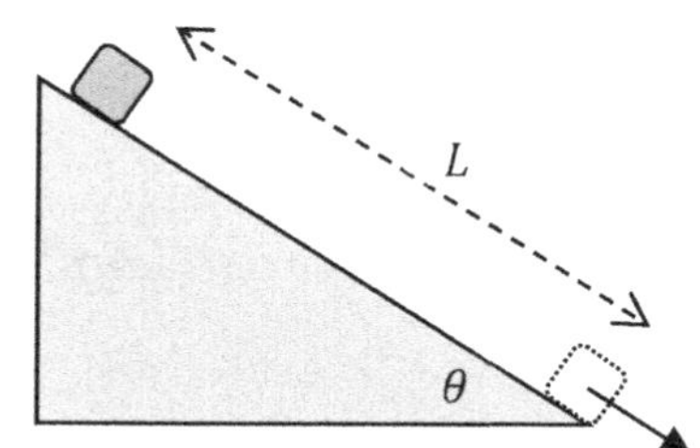

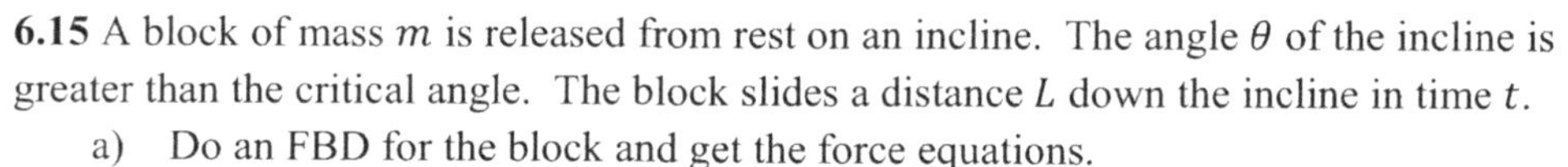

a) Do an FBD for the block and get the force equations.
b) Determine the coefficient of friction in terms of m, g, L, t, and θ. Hint: you can use kinematics to eliminate a! Note: some givens may not be necessary…

6.16 Mass m_1 is on an incline connected to mass m_2 over a pulley. The system is released from rest when m_2 is a height h above the ground. The coefficients of friction between m_1 and the incline are μ_s & μ_k. **Assume m_2 moves towards the earth.**

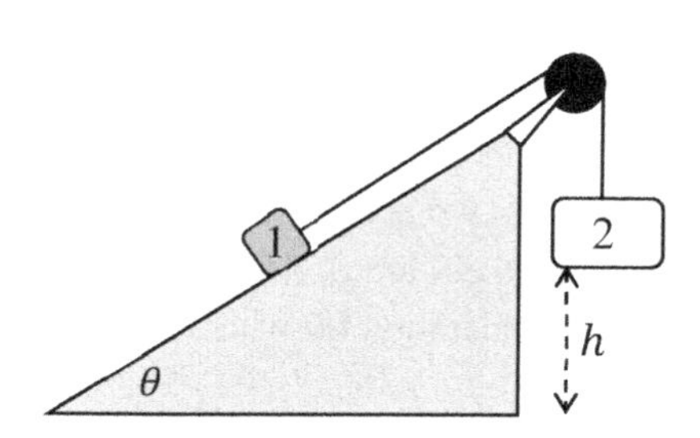

a) What conditions apply if the tension on each side of the pulley is equal?
b) What condition applies if the acceleration magnitude of each block is equal?
c) What condition applies if the acceleration is to be constant?
d) How should the tension magnitude T compare to $m_2 g$?
e) Draw FBDs for each block and determine the force equations.
f) Determine the impact speed of m_2.
g) What is the <u>minimum</u> m_2 required to cause m_1 to slide upwards?

6.17 If this gives you trouble, try **6.18½** first. Two blocks (mass $2m$ and m) are at rest on an incline with a slight angle θ. Frictional coefficients are $\mu_s = 0.50$ & $\mu_k = 0.30$ exist between each block and the incline while there is negligible friction between the blocks. A zombie applies a horizontal force to $2m$.

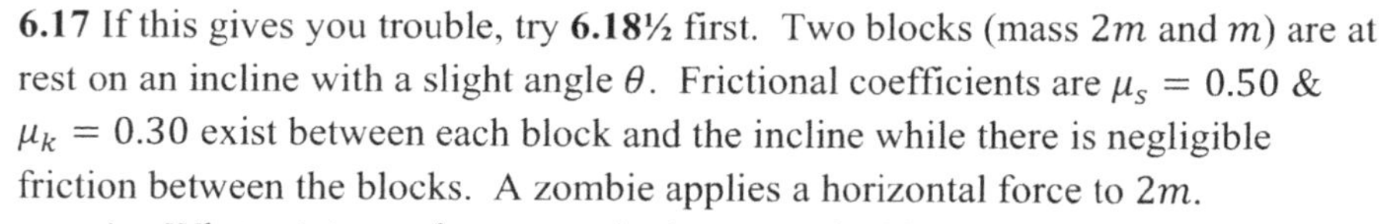

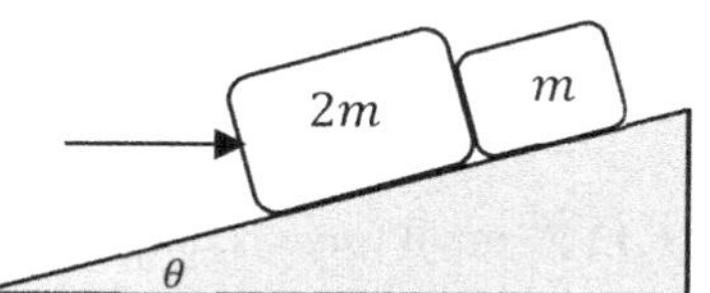

a) What *minimum* force magnitude causes the blocks to slide *up* the ramp? Said another way, what is *largest* force magnitude that can be applied *without* the blocks slipping?
b) What normal force magnitude acts *between the blocks* when they are on the verge of slipping up the ramp?

6.18 Suppose the angle in the previous problem was very large instead of very small. In such a scenario the blocks would tend to slide *down* the plane regardless of the applied force magnitude. Suppose you then wanted to determine the minimum force required to prevent the blocks from sliding *down*.

a) What things would change in the previous FBDs and force equations?
b) Can you explain why, when the incline is greater than a certain angle, no matter how hard you push you can never make the blocks go up the ramp?
c) **Challenge:** determine what factors determine the limiting incline angle described in part b. Do the masses of the blocks matter or not? Come up with an equation that helps one determine this limiting angle for an arbitrary scenario.

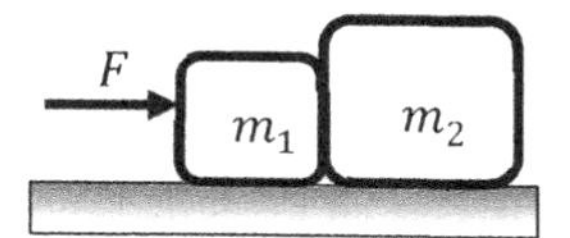

6.18½ Consider two blocks (masses m_1 & m_2) accelerating to the right on a horizontal surface. The blocks have respectively. Force magnitude F is applied horizontally to m_1. Frictional coefficients between each block and the surface are μ_s & μ_k.

a) Under what circumstances is it appropriate to use a *system* FBD for blocks 1 & 2?
b) Determine the acceleration magnitude.
c) Determine the normal force (magnitude) acting between the blocks.
d) Assume $m_1 = 1.00$ kg, $m_2 = 2.00$ kg, $F = 40.0$ N, $\mu_s = 0.777$ & $\mu_k = 0.666$.
   Determine numerical values for acceleration magnitude and normal force (magnitude) between the blocks.
e) Determine the *minimum* applied force magnitude required to make the blocks move.

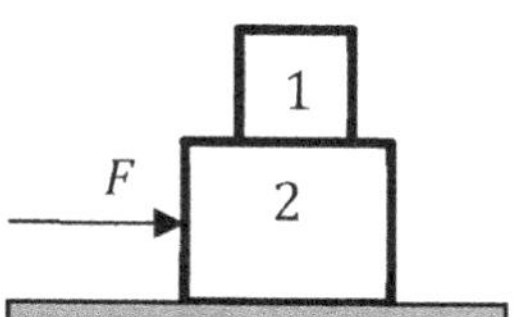

6.19 A zombie pushes on some blocks. Starting from rest, the bottom block is pushed horizontally by a force with magnitude F and accelerates to the right. The frictional coefficients between all surfaces are $\mu_s = 0.600$ & $\mu_k = 0.500$. The two blocks are observed to move together as a single unit (m_1 does not slide relative to m_2).

a) Draw FBDs (and write force equations) for block 1, block 2, & the two-block system.
   Assume f_{12} is the frictional force between 1 & 2, n_{12} is the normal force between 1 & 2, f is the frictional force between the floor and 2, and n is the normal force between the floor and 2.
b) Now you are told $m_1 = 1.0$ kg, $m_2 = 2.0$ kg, and $a = 5.5\ \frac{\text{m}}{\text{s}^2}$. Let $g = 10\ \frac{\text{m}}{\text{s}^2}$. Determine f_{12} & F.
c) Using the same masses, determine the maximum possible F for which the blocks move as a single unit.
d) Write sentences describing the action-reaction pairs for all forces acting on 2. For each pair include the objects involved, the type of force, and the directions of each force in the pair.

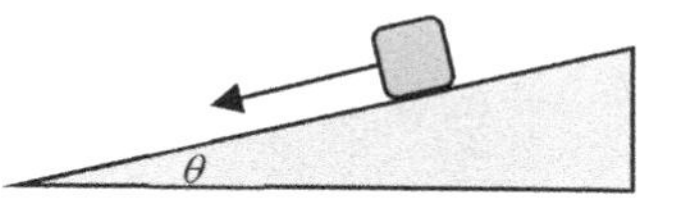

6.20 A box of mass m is placed on an incline. Between the box and the incline $\mu_s = 0.900$ and $\mu_k = 0.800$. The invisible woman pulls on the block with force magnitude F directed down the plane (parallel to the plane's surface).

a) First, *ignore the applied force* and determine the critical angle (angle at which block slips).
b) Assume the angle is $10.0°$. What minimum force magnitude causes the block to slide (in terms of mg)?
c) Now suppose the invisible woman pulled *up the plane* instead of down.
   What minimum force magnitude is required make the block slids up the plane? Is it the same as part b?

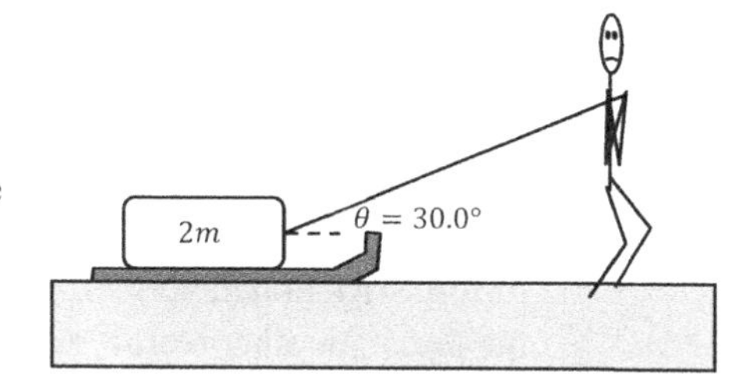

6.21 A zombie pulls a box to the right using a light, inextensible string. The box is on a sled. The mass of the sled is m and the mass of the box is $2m$. The mass of the zombie is $4m$. The box does not slide relative to the sled. There is essentially no friction between the sled and the ground. Between the box and the sled $\mu_s = 0.900$ and $\mu_k = 0.800$. The zombie accelerates to the right with $a = g/5$.

a) Determine the tension (magnitude) in the cable.
b) Determine the magnitude of the frictional force acting on the box.
c) Determine the magnitude of the normal force acting between the block and the sled.
d) Compare the frictional force between the block and sled to the maximum possible value.
e) What horizontal force must be exerted on the zombie by the ground to make this situation happen?
f) Which forces are internal to the box-sled system?
g) What is the action-reaction pair associated with the weight of the box?

6.22 Two blocks (mass m & $3m$) are connected by a massless, inextensible string. They are released from rest. The angle is $\theta = 35.0°$.
Between the larger block and the incline $\mu_s = 0.500$ and $\mu_k = 0.450$.
All other surfaces are essentially frictionless.
Determine the acceleration (magnitude) of the blocks and the tension (magnitude) in the string.
Note: as a bonus, show the blocks accelerate for all angles greater than 20.5°.

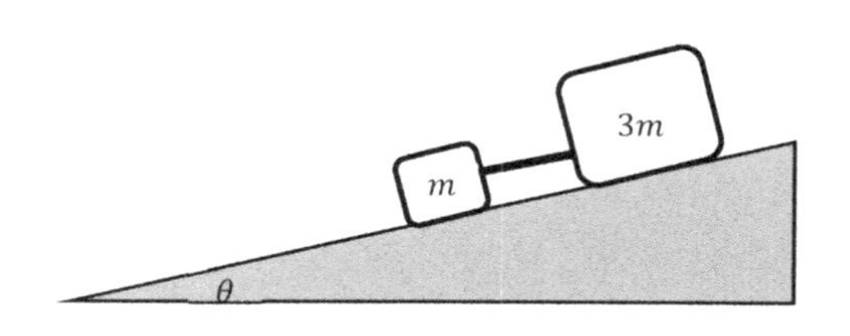

6.23 A box of mass m is *lowered* on an incline at an *increasing rate*. The invisible woman pushes on the block with force magnitude F. Between the box and the incline $\mu_s = 0.900$ and $\mu_k = 0.800$. **I don't care about solving for anything…this is FBD and force equation practice.**

a) Sketch the FBD and write the force equations.
b) What, if anything, would change for your previous results if the block was lowered at a *decreasing* rate?

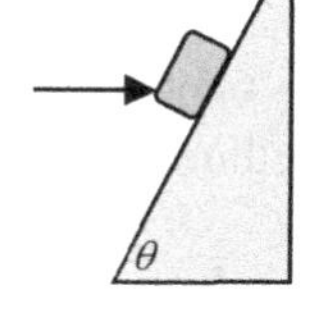

6.24 The figure at right shows masses $4m$ and m which accelerate together on a frictionless surface. The acceleration is caused by a thruster which exerts a constant thrust force T. The coefficient of friction between the two blocks is μ. The magnitude of the acceleration due to gravity is g. This problem only considers the first few seconds of thrust when air resistance may be considered negligible.

a) Determine the minimum thrust magnitude, T_{min}, required to prevent the block of mass m from sliding downward? Answer in terms of m, g, and μ.
b) Now consider the frictional force between the two blocks. If actual thrust force used is twice as big as T_{min}, how is the frictional force on m affected?

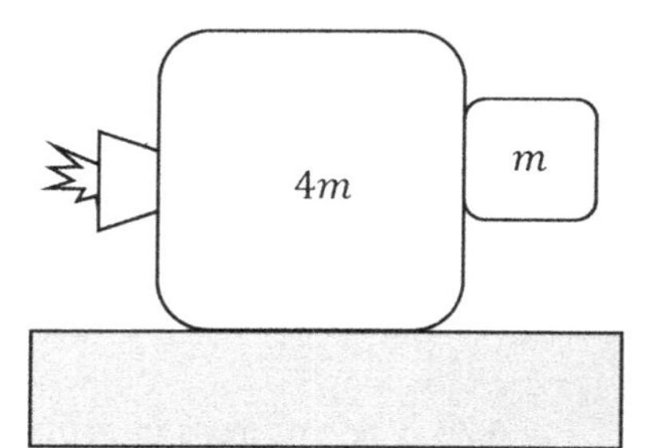

6.25 A board is resting with one end on the ground and the other end help up by an invisible woman. As a result, the board is angled at some angle θ as shown in the figure. A block of mass m is placed on the board and is observed to remain at rest. The coefficients of friction between the block and the incline are μ_s & μ_k. Three forces act on the block: weight, normal force, and friction.

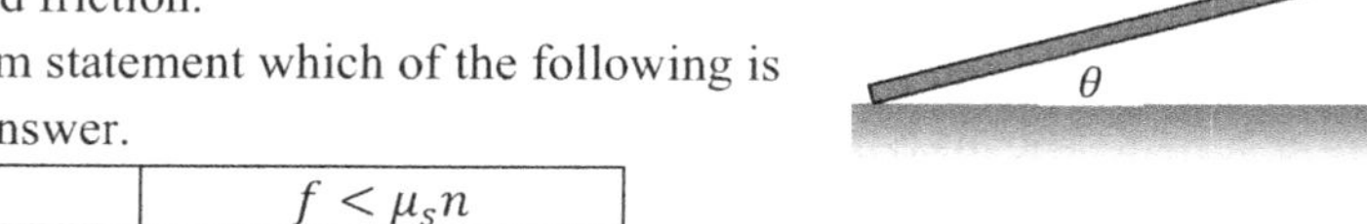

a) According to the wording of the problem statement which of the following is most likely to be true. Circle the best answer.

$f = \mu_k n$	$f = \mu_s n$	$f < \mu_s n$

b) We often discuss action-reaction pairs. Consider the weight of the block as the "action" force. What is the corresponding "reaction" force? Answer by stating 1) the object exerting the force, 2) the type of force (frictional, normal, gravitational, etc), 3) the direction of the force, and 4) the object experiencing the force. In other words, explain the reaction force by filling in the blanks in the following sentence..

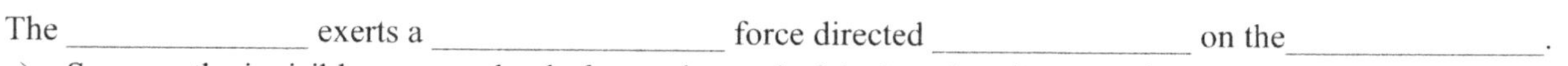
The ______________ exerts a _________________ force directed ________________ on the________________.

c) Suppose the invisible woman slowly *lowers* her end of the board to the ground. How do the following quantities change as the board is *lowered*? Circle the best answer for each case.

Weight	Increases	Decreases	Remains constant
***Component* of weight <u>parallel</u> to the incline**	Increases	Decreases	Remains constant
***Component* of weight <u>perpendicular</u> to the incline**	Increases	Decreases	Remains constant
Normal force	Increases	Decreases	Remains constant
Frictional force	Increases	Decreases	Remains constant
***Max possible* frictional force**	Increases	Decreases	Remains constant

Friction has so many cases it is instructive to create plots. Plots quickly show the value of friction (or some other parameter) for a wide variety of situations. Seeing plots is not nearly as instructive as creating plots yourself. To make the plots you must understand the transition point from static to kinetic friction well. I believe this numerical work should improve your ability to correctly model friction in both FBDs and concept questions.

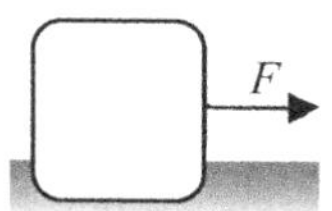

6.26 Friction Graph #1 This is identical to first question in the chapter. Might be worth doing again at this time. A block of mass $m = 1.00$ kg is at rest on a flat surface. The coefficients of friction are $\mu_s = 0.5$ & $\mu_k = 0.3$. A force with magnitude F is applied horizontally to the block. The force is gradually increased from 0 to 10.0 N. Our goal is to plot 1) magnitude of the friction (f) versus F and 2) acceleration magnitude (a) versus F.

e) Determine the minimum applied force (F_{min}) required to make the block slip.
f) Do an FBD assuming the applied force is below F_{min}. Determine equations for both f and a.
g) Do an FBD assuming the applied force is above F_{min}. Determine equations for both f and a.
h) Create a table of values showing F, f, and a. Plot f vs F and a vs F.
i) If a different mass was used, how would your plots change? Explain.
j) Suppose we found out after the fact that $\mu_s = 0.4$ but the other frictional coefficient was still $\mu_k = 0.3$. How would this correction affect the graphs?
k) I created my plots in Excel. I created equations based on the values of $\mu_s$and μ_k shown at the top of the data. Said another way, $\mu_s$and μ_k were constants in my computer code. If I want to consider other frictional coefficients, why can I not simply change the values of my constants to get the new plots?

As a challenge, you could use an IF function in Excel circumvent the issue discussed in part g). It is also great practice to make these plots in MATLAB instead.

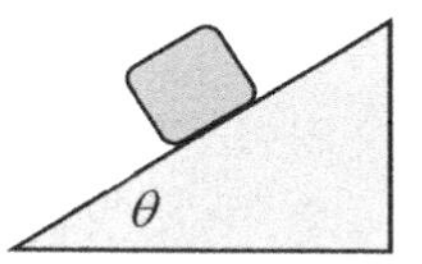

6.27 Friction Graph #2 A block of mass $m = 1.00$ kg is at rest on a ramp. Initially the ramp is initially level with the ground (ramp angle $\theta = 0°$). The coefficients of friction are $\mu_s = 0.5$ and $\mu_k = 0.3$. The ramp angle is gradually increased from 0 to 90°. Our goal is to plot f versus θ and a versus θ. To be clear, I am assuming f is the magnitude of the frictional force and a is the magnitude of the acceleration.

a) Determine the minimum angle θ_c required to make the block slip.
b) Do an FBD assuming $\theta < \theta_c$. Determine equations for both f & a.
c) Do an FBD assuming $\theta > \theta_c$. Determine equations for both f & a.
d) Create a table of values showing θ, f, and a in Excel. Use equations in Excel to reduce the amount of computation. Plot f vs θ and a vs θ.
e) If a different mass was used, how would your plots change? Explain.
f) Suppose we found out after the fact that $\mu_s = 0.4$ but the other frictional coefficient was still $\mu_k = 0.3$. How would this correction affect the graphs?

Hint: Excel assumes the arguments of the cosine and sine functions are in radians. To convert degrees to radians, use the Excel function RADIANS.

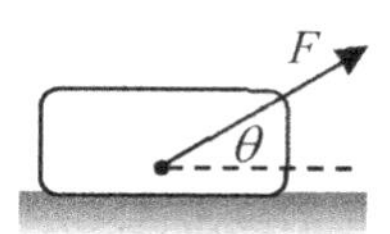

6.28 Friction Graph #3 A block of mass $m = 1.00$ kg is at rest on a flat surface. The coefficients of friction are $\mu_s = 0.5$ and $\mu_k = 0.3$. A constant force with magnitude $F = 8.00$ N is symmetrically applied to the block just below its center of mass. By applying the force in this manner, the block lies flat without twisting on the surface as the angle θ is gradually increased from 0°. In this problem f means the magnitude of the frictional force and a is the magnitude of the acceleration.

a) **For this problem, $F < mg$.** From this info we know the block will remain in contact with the table.
If we instead used $F > mg$, the block lifts off the table if $F \sin\theta > mg$.
If we instead used $F > mg$, our FBD makes no sense for angles greater than $\theta_{max} = \sin^{-1}\left(\frac{mg}{F}\right)$.

b) Show the block is on the verge of slipping when $\theta_c = 83.3°$. Hints are given in the solutions.

c) For $\theta < \theta_c$, determine equations for f and a.

d) For $\theta_c < \theta < \theta_{max}$, determine equations for f and a.

e) Create a table of values showing θ, f, and a. Plot f vs θ and a vs θ.

f) Check your plot by using a derivative to determine the angle at which acceleration is maximized. Notice this implies it is easier to drag an object by pulling at a slight angle rather than pulling horizontally!

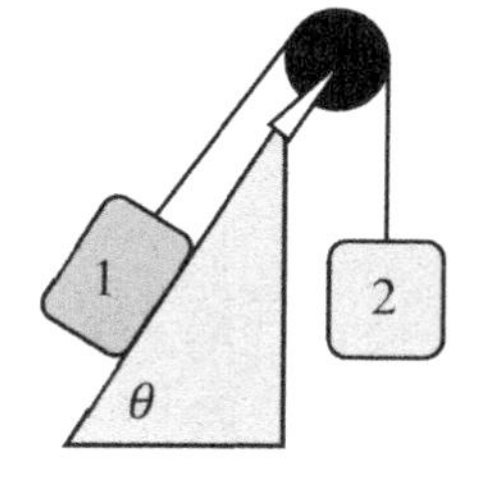

6.29 Friction Graph #4 A two block system is designed as shown. Block 1 has mass $m_1 = 1.00$ kg and experiences coefficients of friction $\mu_s = 0.5$ and $\mu_k = 0.3$. The ramp has fixed angle $\theta = 60.0°$. A student in lab varies m_2 from 0 to 5.00 kg. For each value of m_2 the system is released from rest. For all cases in this problem assume up the ramp the positive direction for m_1.

a) What value of m_2 balances the system with zero frictional force? Note: when m_2 is larger than this value, friction points down the plane. For m_2 less than this value friction points up the plane.

b) Determine the smallest hanging mass (m_{2min}) that keeps the system at rest.

c) Determine the largest hanging mass (m_{2max}) that keeps the system at rest.

d) For $m_2 < m_{2min}$, use FBDs to determine equations for the acceleration and friction. Remember, most exam questions will not have numbers for you to plug in. Before plugging in any numbers, first find your equations in terms of m_1, m_2, θ, g, and μ_k.

e) For $m_2 > m_{2max}$, use FBDs to determine equations for the acceleration and friction. First find your equations in terms of m_1, m_2, θ, g, and μ_k; then plug in numbers.

f) For $m_{2min} < m_2 < m_{2max}$, use FBDs to determine equations for the acceleration and friction. First find your equations in terms of m_1, m_2, θ, and g; then plug in numbers.

g) With complicated equations, it is nice to have as many checks as possible. Checking the units is a great start but consider also checking if your equations give sensible results in the following cases:
   i. $m_2 \to \infty$ (m_2 approximately in free fall)
   ii. $\mu_k = 0$ (if no friction does $m_2 = m_1 \sin\theta$ balance the system)
   iii. $\theta = 90°$ (Atwood's machine, friction and normal force should drop out)

h) Make a table of m_2, $\vec{f}$, and $\vec{a}$. Plot $\vec{f}$ vs m_2 and $\vec{a}$ vs m_2. Remember that we have chosen up the incline as the positive direction for all cases. We are plotting the vectors this time (not the magnitudes).

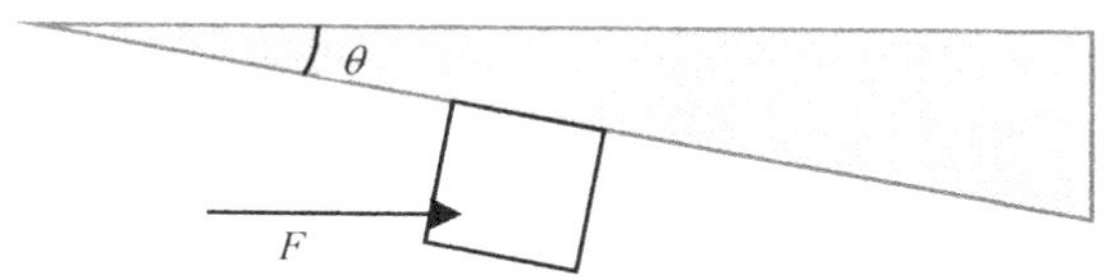

6.30 A block is pushed by a force F along a ceiling as shown in the figure. The block has mass m. The coefficient of kinetic friction between the block and the ceiling is μ. Assume that the block never loses contact with the ceiling.

a) Determine the minimum force magnitude required to keep the block in contact with the ceiling.
b) Determine the acceleration (magnitude) of the block assuming the applied force has magnitude greater than the min value determined in the previous part.
c) To get a feel for how ridiculous this problem is, assume $\theta = 15°$, $\mu = 0.5$, and $m = 0.1$ kg. Perhaps this would model an eraser going across an angled ceiling? What is a numerical value of the minimum applied force magnitude (and corresponding acceleration)? Compare the applied force magnitude to the object's weight and the acceleration to g.

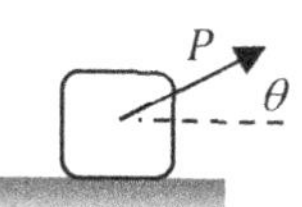

6.31 A block is pulled along a horizontal floor with an angled rope as shown in the figure. The coefficients of friction between the block and the floor are μ_k & μ_s. Assume the force P is applied at the center of mass so that we need not worry about one side of the block lifting up.

a) For a given angle, what is the largest possible P before the block lifts off the ground?
b) Show the force that puts the block on the verge of slipping is $P_{min} = mg\frac{\mu_s}{\cos\theta+\mu_s\sin\theta}$.
c) Consider some interesting limiting cases of your above answers to see if they make sense. What if there was no friction? What happens to the equations for 0° and 90°?
d) Determine the acceleration of the block for $P > P_{min}$.
e) **Requires Calculus:** For a fixed value of $P > P_{min}$, what angle gives the maximum acceleration? This is the optimum angle to pull on the pull on the block if you want to drag it with a minimal amount of effort.

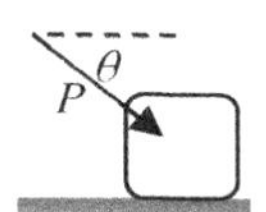

6.32 A block of mass m is at rest. The block is pushed by a force with magnitude P as shown in the figure. The coefficients of friction between the block and the floor are μ_k & μ_s.

a) Suppose the block is on the verge of slipping. Determine μ_s.
b) Consider your results for 0° and 90°. Does the model make sense?
c) Consider the results as $P \to 0$ (or $m \to \infty$). Does the model make sense?
d) Now suppose the block is accelerating. Determine the acceleration in terms of P, m, g, θ and μ_k.
e) **Requires Calculus:** What angle gives the maximum acceleration?
f) Compare your results from parts d) and e) to **6.31d** and **6.31e**. Consider what should happen if you shift the angles of **6.31** from θ to $-\theta$. Note the trig identities $\sin(-\theta) = -\sin\theta$ while $\cos(-\theta) = \cos\theta$.

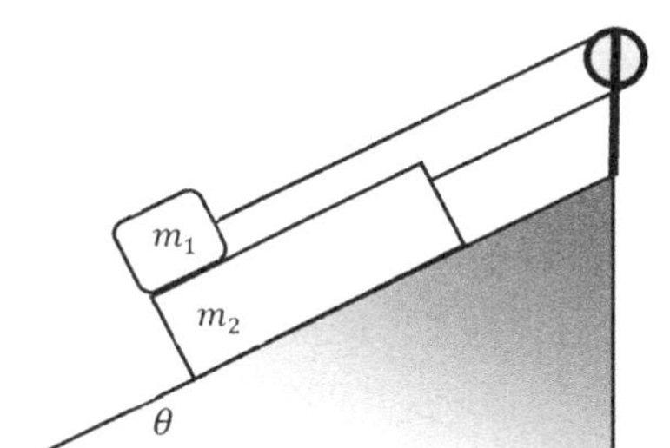

6.33 Block 1 rests on block 2 which rests on an inclined plane as shown. The blocks have masses $m_1 = m$ and $m_2 = 4m$. There is no friction between the plane and m_2 but between m_1 and m_2 the coefficients of friction are $\mu_k = 0.10$ & $\mu_s = 0.15$. The blocks are connected by a massless, inextensible string over a massless, frictionless pulley.

a) Assuming θ is large, which direction would you expect block 2 to move?
b) Determine the minimum angle to cause the system to slide.
c) Assume the angle is now 60.0°. Determine the acceleration of the blocks and the tension in the string.
d) **Challenge:** Assume the system starts form rest and the length of m_2 is $L = 1.00$ m. Determine the time required for m_1 to slide off the top of m_2. Assume the size/length of m_1 is small compared to m_2.

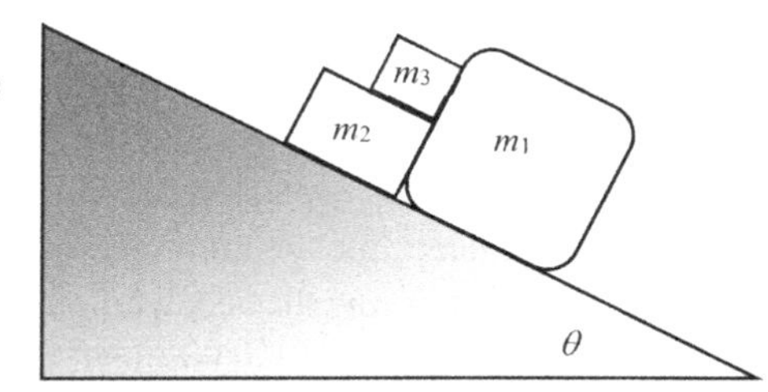

6.34 Do the FBDs for each block separately and for all three as a system. Assume the static and kinetic coefficients of friction between m_1 and the incline are μ. **Assume there is negligible friction on all other surfaces to keep the problem manageable.** The system is at rest but just about to slip down.

a) Determine critical angle of the system in terms of the masses and μ.
b) Determine all normal forces (magnitudes) in the problem.
c) Think: for angles above the critical angle, which of your previous normal forces are affected?

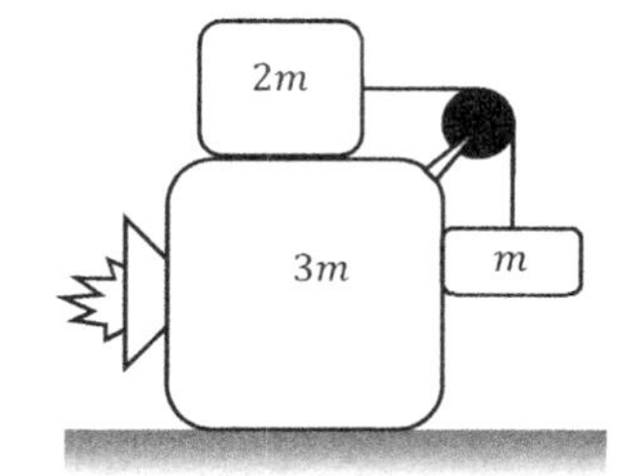

6.35 Block 3 is accelerating to the right due to a constant thrust force with magnitude F. Blocks 1 and 2 are connected using a light, inextensible string over a pulley with negligible axle friction. To be clear, $3m$ includes everything except the string, m, and $2m$. Assume friction is negligible everywhere except between block 3 and the ground with coefficient $\mu_k = 0.25$. Assume the thrust force is just right so blocks $2m$ and m do not move relative to block $3m$.

Note: in this problem the pulley has mass. *In general*, if the pulley has mass, we cannot assume the tension in the string is the same on both sides. HOWEVER, *if the pulley is not rotating*, we may still assume the tension in the string is the same for both m and $2m$. More on this in Chapter 10!

a) Detemine the tension (magnitude) in the cable.
b) Determine the acceleration (magnitude) required to keep block 2 from sliding relative to block 3.
c) Determine the thurst force (magnitude).
d) Assuming the pulley (black circle only) has mass $m_p = m/5$. Determine the magnitude and direction of the force exerted on the pulley by the the pulley support (white triangle).
e) **Challenge:** Now assume the coefficents of friction between all surfaces are $\mu_s = \mu_k = 0.25$. Detemine the range of acclerations possible to keep block 2 from sliding.

6.36 Challenge: In-depth Table Cloth Trick

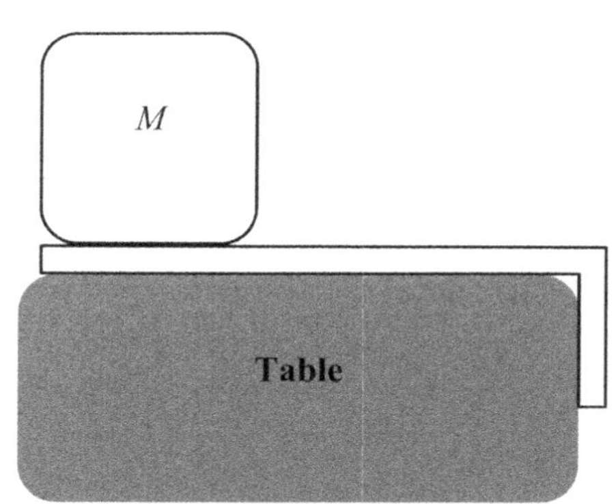

Suppose we have a very smooth tablecloth with mass $m = 0.100$ kg. On top of the table cloth, sits block of mass $M = 1.00$ kg. A cube is placed at the edge of the table cloth and has side $s = 0.20$ m. Assume the coefficients of friction between all surfaces are $\mu_s = 0.5$ and $\mu_k = 0.3$.

a) What minimum force (magnitude) must be applied to make the block on the verge of slipping relative to the tablecloth? Assume the force is applied horizontally to the right.
b) Suppose the applied force is 45 N (about 10 lbs). What is the acceleration (magnitude) of the block (relative to the earth)?
c) What is the acceleration of the tablecloth (relative to the earth)?
d) How long will it take for the block to slide off the tablecloth completely? For simplicity, that the friction force is constant even while the block is partially off of the tablecloth.
e) Determine the speed of M at the instant the tablecloth is completely removed beneath it.
f) Relative to the table, how far will block M move as the table cloth is jerked out underneath it?

Notice the friction force on M is the same size regardless of the size of the applied force. Using the large force reduces the time that friction acts on M. This in turn means M will not pick up appreciable speed and will not move a noticeable amount. The block M will appear to stay approximately in place while in reality it does move a little bit in the same direction of the table cloth.

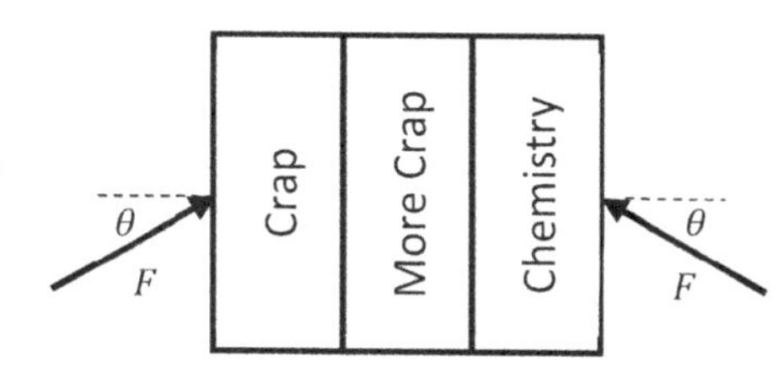

6.37 Suppose you are holding three books by pressing them together. Your hands exert a net force with magnitude F angled as shown. Note: this $\vec{F}$ includes the normal and frictional forces from your hand; it is the *net* force from each hand on the book. Coefficients of friction between the books are $\mu_s = 0.5$ and $\mu_k = 0.3$. The mass of each book is m.

a) Determine the minimum force magnitude F_{min} required to prevent the middle book from slipping out. Check your result as $\theta \to 0°$ & $\theta \to 90°$.

b) In real life a zombie applies *more* than this minimum force to ensure the books won't slip. What must happen to θ if you apply $F > F_{min}$? How does the frictional force between the books change?

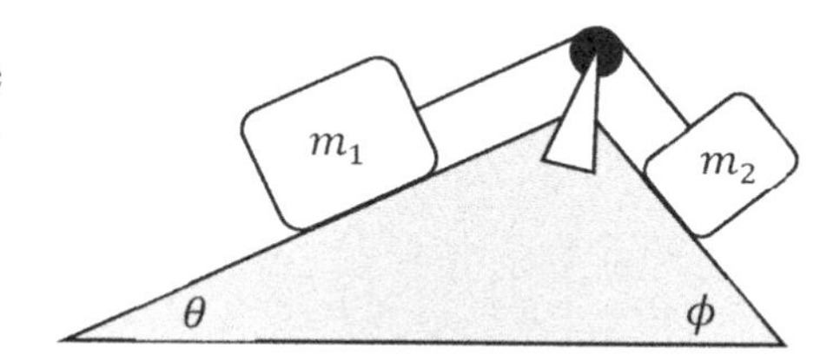

6.38 Suppose two blocks are sitting on two different inclines. The blocks are attached by a string of negligible mass over a pulley with negligible axle friction. The coefficients of friction between the blocks and the inclines are $\mu_{s1} = 0.15$ and $\mu_{k1} = 0.10$ and $\mu_{s2} = 0.25$ and $\mu_{k2} = 0.20$ respectively. The angles are θ and ϕ while the masses are m_1 and m_2.

a) Think: before grinding through a massive calculation with tons of different coefficients, how could you first figure out which direction friction will point on each block?

b) What steps would you use to determine if the blocks slides or remain stationary?

6.39 A box of mass m sits on a flat bed truck. The truck accelerates from rest with magnitude $a = \frac{g}{5}$. At this rate of acceleration the box doesn't slide off the back of the truck bed. The coefficients of friction between the block and truck are $\mu_s = 0.5$ & $\mu_k = 0.3$.

a) Determine the frictional force (magnitude) on the box.

b) What is the largest possible acceleration the truck could have before the box slides off the back of the truck?

6.40-46 Consider doing some simulations. Some suggested simulations are found in a free worksheet on www.robjorstad.com.

FORCES IN CIRCULAR MOTION

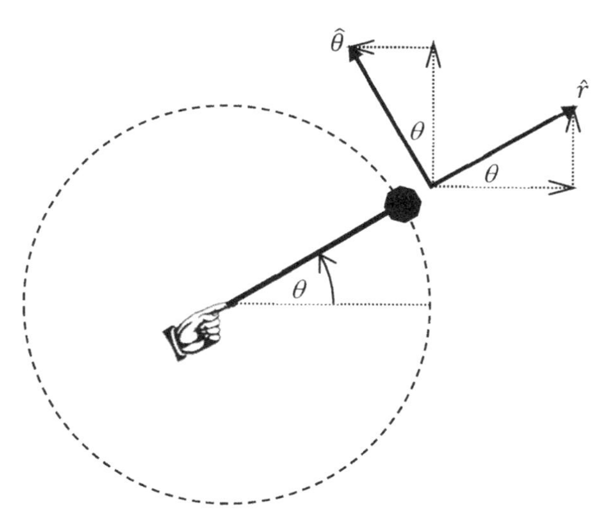

$\hat{r}$ = Shorthand notation for "radially outwards". Unless otherwise specified, when someone says radial direction you are to assume they mean radially outwards.

$\hat{\theta}$ = Tangential direction – a direction tangent to the circle. Note: in math classes they always choose CCW. Sometimes, if everything in a physics problem moves CW, you might flip this direction just as we sometimes called downwards the positive direction.

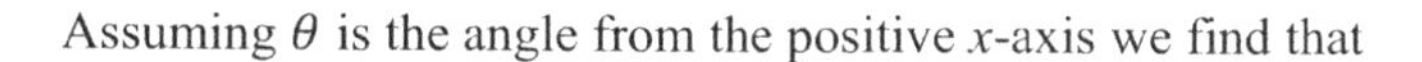

Assuming θ is the angle from the positive x-axis we find that

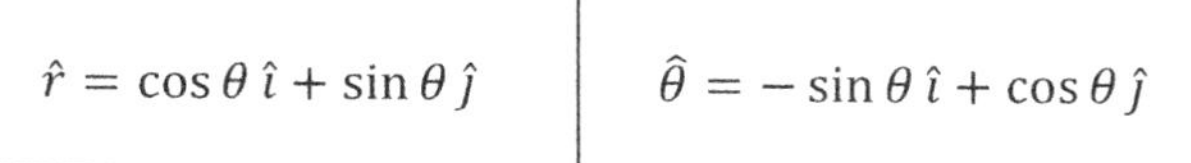

$\hat{r} = \cos\theta\,\hat{\imath} + \sin\theta\,\hat{\jmath}$	$\hat{\theta} = -\sin\theta\,\hat{\imath} + \cos\theta\,\hat{\jmath}$

Correctly label the components of $\hat{r}$ and $\hat{\theta}$ with $\sin\theta$ or $\cos\theta$ as appropriate in the figure.

Some definitions for circular motion <u>with constant radius</u>:

CW – This is usually shorthand for clockwise.

CCW – Similarly, this is shorthand for counter-clockwise.

$\mathbb{T}$ = Period = time to complete a single revolution. To distinguish it from the other t's (time, tension, torque, etc) I try to write this differently than a normal capital T.

v_{tan} = tangential speed. This is the speed of the ball at any given point in the circular motion.

a_{tan} = tangential acceleration. This relates to the rate of change in the speed.

a_r= radial acceleration. This relates to the change in direction of the velocity. Negative numbers for a_r indicate acceleration is towards the center.

a_c = centripetal acceleration. This is the magnitude of a_r.

$|a_c| = |a_r| = \frac{v^2}{r}$. Here v is tangential speed v_{tan} and r is the radius of the circular motion.

Note: since a_{tan} relates to the change in *speed* (not the change in *velocity vector*) we may write

$$a_{tan} = \frac{dv}{dt}$$

Please note that this is not the same thing as

$$\vec{a} = \frac{d\vec{v}}{dt} = a_r\hat{r} + a_{tan}\hat{\theta}$$

In circular motion we know that the radial component is actually inwards which means the minus sign is implicit in ar. To make it explicit one often rewrites the above equation as

$$\vec{a} = \frac{d\vec{v}}{dt} = a_c(-\hat{r}) + a_{tan}\hat{\theta}$$

See the difference? Watch out for a_c versus a_r! Remember, they both have magnitude $\frac{v^2}{r}$.

Which way does acceleration point during circular motion?

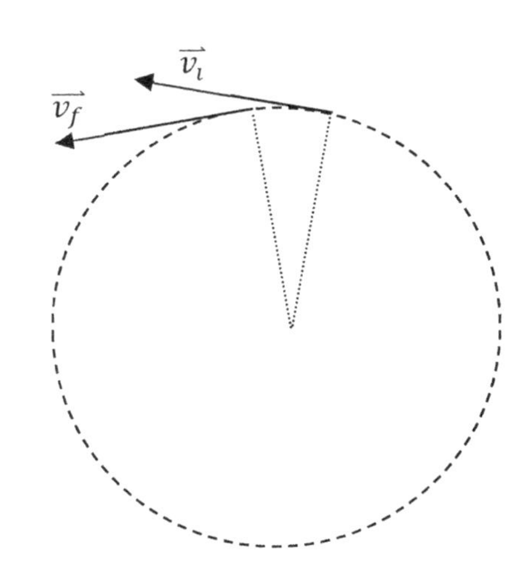

6.47 Consider an object that is going counter-clockwise along a circular path as shown in the figure at right. In the figure I have drawn the velocity vector just before and after it reaches the top of the circle. Notice this object is moving with constant speed since the length of each velocity vector is the same.

a) Graphically (tail-to-tip) figure out the direction of the acceleration at the top of the circle. Hint: the acceleration vector is given approximately by

$$\vec{a} \approx \frac{\Delta\vec{v}}{\Delta t}$$

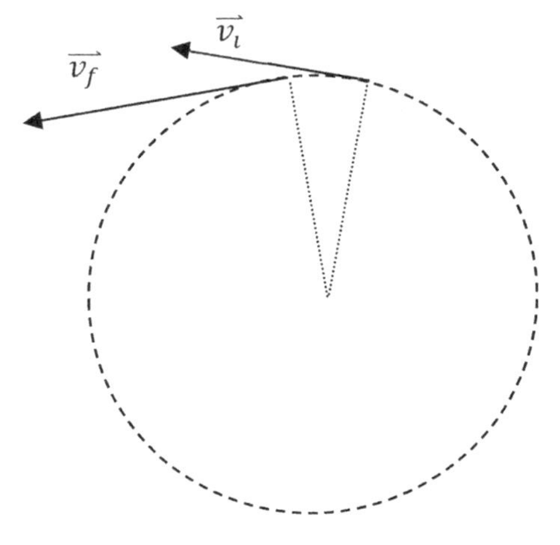

6.48 Consider the case where the final speed is 30% greater than the initial.

a) Graphically determine the acceleration direction at the top of the circle.
b) Split that acceleration into a component towards the center of the circle and a component tangent to the circle.

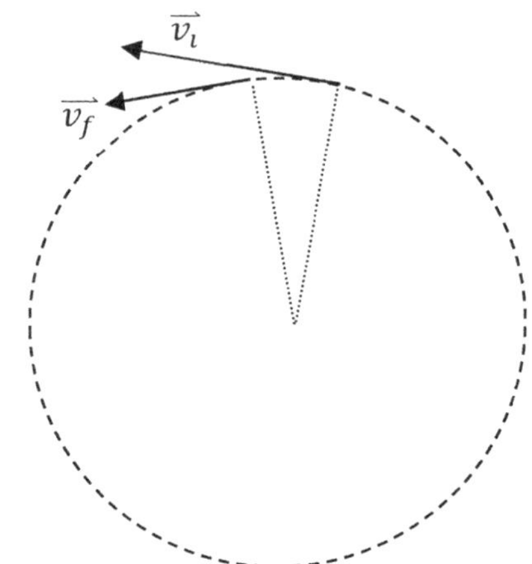

6.49 Now consider the case where the final speed is 30% less than the initial.

a) Graphically determine the acceleration direction at the top of the circle.
b) Split that acceleration into a component towards the center of the circle and a component tangent to the circle.

6.50 Think Quickly! In a car there are two ways to change your speed but three ways to accelerate. What are they? Specifically, how could you maintain constant speed while changing your velocity?

6.51 A ball starts from rest at point **A**. It swings on a rope all the way over to point **E** where it momentarily comes to rest. It then swings back to point **A**.

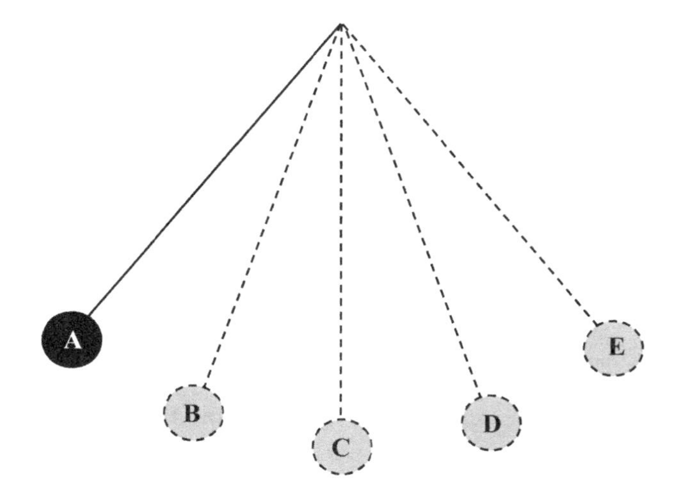

a) Sketch and label a_{tan} (direction AND size of the arrow) at all five points. Can it be zero?

b) Sketch and label the a_c (direction AND size of the arrow) at all five points. Can it be zero?

c) At which point(s), if any, is the <u>total</u> acceleration zero?

d) For the initial swing from **A** to **E** determine where the ball is speeding up, slowing down or at rest.

e) **Consider doing a simulation question right after this?**

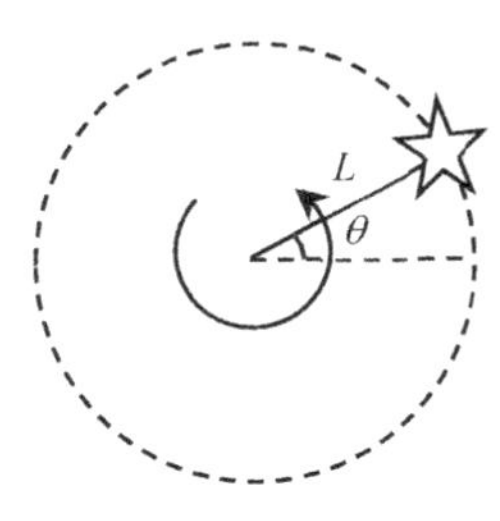

6.52 The star of mass m moves in a vertical circle on a fixed length L of string. The string has negligible mass. Assume the mass is 0.200 kg. The distance from the center of the circle to the center of the star is 1.00 m. The size of the mass is small compared to the length of the string. At the instant shown, $\theta = 30°$ and $v = 5.00\,\frac{\text{m}}{\text{s}}$.

a) Draw the FBD for the star at the instant shown in the figure. Use radial/tangential coordinates OR in centripetal/tangential coordinates. Don't plug in any numbers yet so you can check your work with the solutions.

b) Determine the magnitude of the tangential acceleration.

c) Determine the magnitude of the centripetal acceleration.

d) Sketch a picture of the total acceleration vector similar to the one shown at right below the circle. On your sketch clearly label the magnitude and direction of $\vec{a}_{total}$.

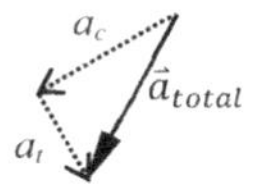

6.53 Uniform circular motion: Relating RPMs, Period, and speed

Uniform circular motion is a fancy way of saying circular motion with constant speed.

a) We know the distance covered in one orbit is $2\pi r$. The period ($\mathbb{T}$) is the time to complete one orbit. Determine $\mathbb{T}$ in terms of v and r.

b) The rotation rate ω is called angular frequency (or angular speed). The standard units for ω are rad/s. We will learn later that $\omega = \frac{2\pi}{\mathbb{T}}$. Determine ω in terms of v and r.

c) An amusement park ride rotates at 10 RPMs. The rider has a radius of 3.00 m.

  i. Determine the rotation rate in rad/s. Note: 1 rev = 2π rad.

  ii. Determine the period of the rotation.

  iii. Determine the speed of the rider.

6.54 Consider an amusement park ride such as the one shown at right. The two riders rotate in a vertical circle with constant speed v and radius r.

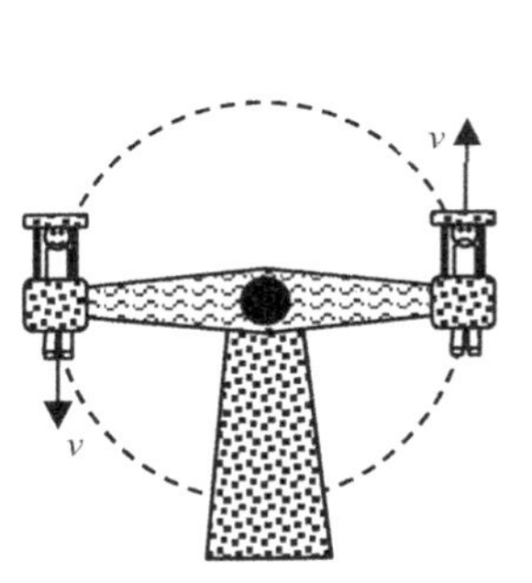

a) Consider the rider on the right. In which direction does the ride/seat exert a force exerted on the rider: up and to the right or up and to the left?

b) Which direction is the total acceleration of the rider on the right?

c) Draw the FBD and determine force equations for the rider on the right.

d) Assume each rider is 65 kg and has radius 5.0 m. What speed is required for the riders to experience an acceleration (magnitude) of $0.4g \approx 4\,\frac{\text{m}}{\text{s}^2}$?

e) Determine the net force exerted by the seat/ride on the rider on the right. Express your result with magnitude and direction as well as in Cartesian form ($\hat{\imath}$ and $\hat{\jmath}$).

f) How do the results change if we instead consider the rider on the left?

6.55 Swing the bucket of water

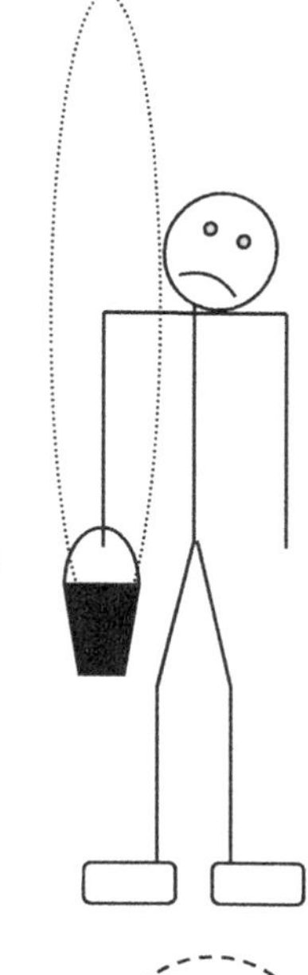

Now imagine you have a bucket with some water in it. You swing the bucket in a vertical circle of radius r. The water in the bucket has mass m. I'm interested in finding out the minimum speed one needs to swing the bucket to prevent the water from spilling out. **To simplify the problem, let's assume that somehow the bucket swings at constant speed v.**

a) Where is the water most likely to fall out? At the top, of course. Draw an FBD and write down the force equation for the water in the bucket at the top.

b) The bucket must remain moving at the top or the water will fall out. We do know as the bucket spins faster and faster there is more tension in your arm (more normal force on the water from the bottom of the bucket). Conversely, if the bucket spins slower and slower, the normal force on the water decreases. What is the smallest possible value this normal force can be? Any smaller than this and the water is no longer touching the bucket and the problem makes no sense!

c) Determine the minimum speed.

d) Finally, determine the period ($\mathbb{T}$) of the orbit at this minimum speed.

e) Think: is your answer to the previous part a minimum period or a maximum period?

6.56 Consider vertical motion in a circle for two kinds of devices. On the left of the figure is the standard ball on a string. On the right in the figure shows a rod with a mass on each end.

a) Why is it possible for one of these situations to rotate with constant speed but not the other? Explain.

b) Why is it reasonable to consider vertical circular motion at high speeds as constant velocity problems?

6.57 Barrel of Fun (aka The Gravitron)

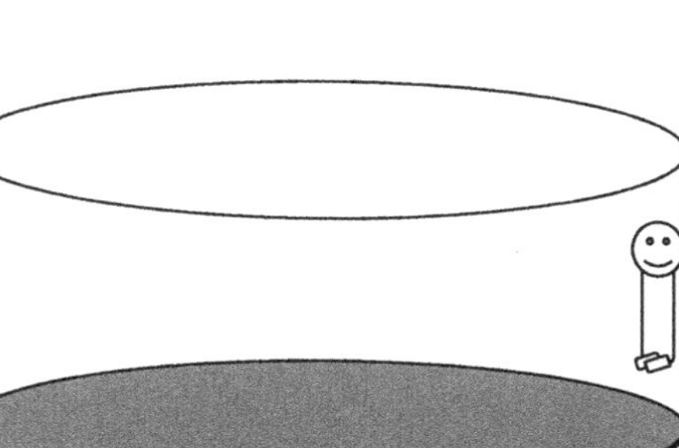

Sometimes at the fair you will see a ride called the Barrel of Fun or The Gravitron. The ride is essentially a cylindrical room that rotates. People stand with their backs against the wall. As the ride gradually spins faster and faster the people press against the wall more and more. Eventually, when the ride reaches some constant max speed, the floor of the room drops down little bit but the people still press against the wall without sliding down!

a) Assume the radius of the room is r. Assume that the coefficients of friction between the person and the wall are μ_s & μ_k. Determine the largest possible period that allows the person to remain suspended above the floor.

b) Determine the magnitude of the force exerted by the wall on the woman at this rotation rate. Express your answer in terms of the appropriate given parameters.

c) The rotation rate doubles. By what factors will the normal force and frictional force change? Explain.

6.58 Which flies off first?

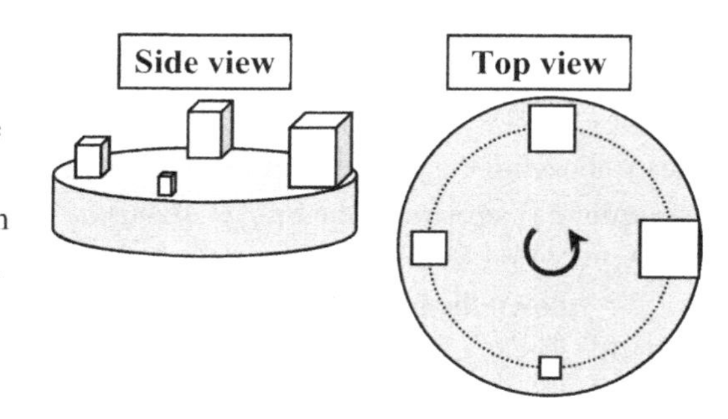

Suppose you have a turntable that spins in the horizontal plane. On the turntable you set a bunch of blocks. You very carefully align the masses so that the center of mass for each block is exactly the same distance from the center of the circle. Here, let us assume once again that $\mu_s = \mu_k = \mu$. You gradually increase the speed of the turntable. Eventually, the masses will come flying off. Which block, the lightest one or the heaviest one, will fly off first? To figure this out, see below:

a) By increasing the speed very gradually, which component of acceleration are we able to ignore?
b) Do an FBD for an arbitrary mass. Use the FBD to determine the period at which the block begins to slip.
c) Does mass affect the period equation? Which object will fly off first?
d) Determine an equation for the coefficient of friction in terms of $\mathbb{T}$, r, and g.

6.59 Car going around a flat curve

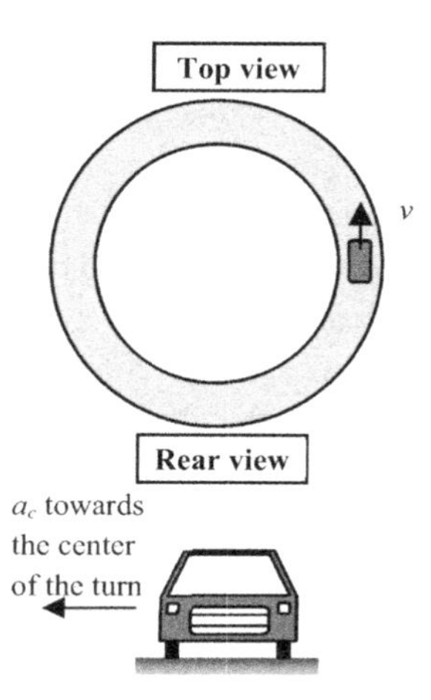

The above problem is very similar to a car driving on a flat road as it navigates a circular turn. Assume the car is moving with constant speed v. The coefficients of friction between the tires and the road are μ_s & μ_k. If the car is going too fast, or if the turn is too tight, the car will *slide* sideways just like a box sliding on plane.

Determine the minimum turn radius for this car at this speed on the flat track.

Note: if you are wondering about the car tipping instead of slipping, wait until you understand torque in a later chapter.

Comments on the frictional forces involved with tires

Rolling friction is mainly caused by the tires being slightly smashed by the road as they turn. As the wheel rotates, a different part of the tires is being smashed every second. This wastes some of the energy that could be used by forward motion. The rolling friction is thus pointed opposite the direction of motion. Please note: rolling friction is *not* given by $f = \mu n$. For the most part we ignore rolling friction in this course. If you want to learn more, try the internet…

When an object rolls without slipping, the point of contact with the ground is not moving relative to the ground. This implies static friction is also keeping the object (say a tire) from spinning relative to the ground. It is possible to have the wheels spin without the car moving forward…but this is not the case of rolling without slipping. This is a "burnout". We will discuss rolling motion after learning about rotation and torque.

If one tries to navigate a turn too quickly, the car will slide sideways. In this case, the sliding is perpendicular to the direction of rotational motion. The frictional force used on the verge of slipping out of the turn is given by $f = \mu_s n$.

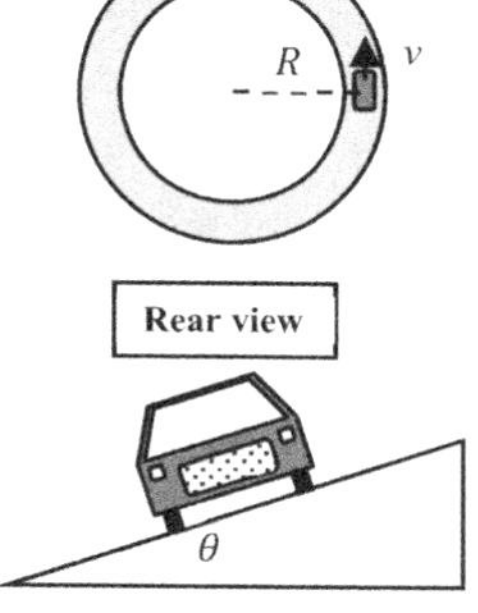

6.60 A car of mass m drives with constant speed on a banked curve on a circular race track of radius R. The angle of the banked curve is θ as shown. Suddenly the car hits a patch of ice. While on the ice, assume negligible friction force act on the car.

a) Do an FBD and write the force equations.
b) Determine the speed required to negotiate the turn without slipping. Let us call this speed the ideal speed for a banked curve. Does the mass matter?
c) Determine the normal force (magnitude) acting on the car at the ideal speed.
d) For speeds below the ideal speed, which way will the car slide: towards the inside or the outside of the turn?

6.61 A car of mass m drives on a banked curve. The figure is similar to the previous problem but now friction does exist between the tires and the road. This time we know the radius of the curve is R and the speed of the car is v.

a) Assume the car is on the verge of slipping to the outside of the curve. Draw the FBD and write down the force equations
b) Determine the minimum coefficient of friction required to keep the car in the turn.
c) Assume the car is on the verge of slipping to the inside of the curve. Determine the minimum coefficient of friction required to keep the car in the turn.
d) Assume the car has speed $v = 30.0\frac{\text{m}}{\text{s}}$, and the radius of the turn is $R = 200.0$ m. Determine the ideal angle to bank the curve. At the ideal angle no friction is required to execute the turn.
e) Create a table and plot of μ_{min} vs θ. For θ below the ideal angle, use the formula you found in part a). For θ above the ideal angle, use part b).

6.62 A car of mass $m = 900.0$ kg drives on a banked curve with angle $\theta = 15.0°$. The figure is similar to the previous problems. The radius of the curve is $R = 200.0$ m and the speed of the car is $v = 20.0\frac{\text{m}}{\text{s}}$. Typically, friction does exist between the tires and the road with coefficients $\mu_s \approx 1$ (rubber on concrete). Unfortunately, today the road is icy. On an icy day the frictional coefficient between the tires and the road drops to about 0.15.

a) Determine the ideal speed for this curve.
b) Does friction point up or down the plane? Hint: compare the car's speed the ideal speed.
c) Determine the normal force (magnitude) acting on the tires.
d) Determine the frictional force (magnitude) acting on the tires.
e) Is it appropriate to assume $f = \mu_s n$ for this problem? Explain.

6.63 A car drives on a banked curve. The figure is similar to the previous problems. Friction exists between the tires and the road with coefficients μ_s & μ_k. The radius of the curve is R and the angle of the incline is θ.

a) Determine the maximum speed for which the car is able to execute the turn.
b) Assume $R = 200$ m & $\mu_s = 0.80$. Create a table and plot θ in degrees vs v_{max} in miles per hour.
c) Assume $R = 200$ m & $\mu_s = 0.15$. Create a table and plot showing θ in degrees vs v_{max} in miles per hour.
d) **Challenge:** when you make the above plots it appears like increasing the coefficient of friction decreases the range of useable angles for a banked curve. Explain why this is not true.

Consider performing an internet search for "banked curve interactive" or "banked curve applet" to learn more online. You may have also heard about the concept of negative lift or downforce in auto racing. I typically ignore this for a 1st year physics course. Negative lift uses the aerodynamics of the vehicle to push air upwards from the car. Since the car pushes air upwards, the air pushes downwards on the car causing the normal force to increase. Since the maximum possible value of friction depends on the normal force, negative lift would modify all of our previous results.

6.64 Sliding over a hill

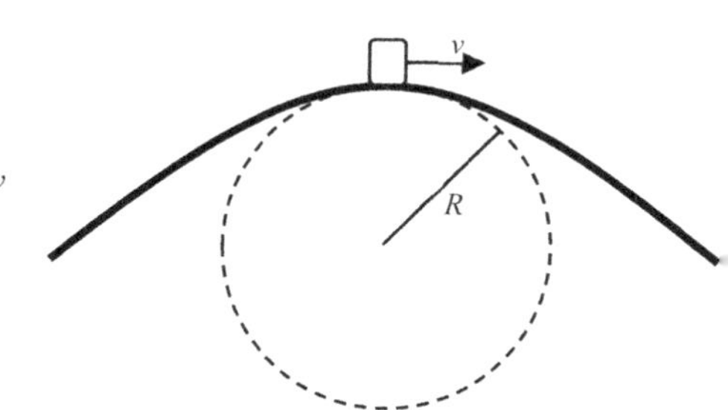

Suppose a block of mass m is sliding over a frictionless hill (we will discuss rolling in a later chapter). The very top of the hill can be modeled as circular motion as shown in the figure. Suppose you are told the speed of the block is v and the radius of curvature of the hill is R. In this case, let's assume that the height of the block is no longer negligible compared to R. In particular, let's assume that the height of the block is $R/4$.

a) Verify that the radius of circular motion in this case is actually $9R/8$. Notice this is a correction of 12.5%.
b) Determine an expression for the normal force (magnitude) acting on the block at the top of the hill in terms of m, g, R, and v. Think about your result: as the speed increases, what should happen to the normal force?
c) What is the *maximum* speed one can drive over the hill without losing contact with the road? Think about your result: if the radius of the hill increases, what should happen to v_{max}?
d) Compare this to the swinging the bucket of water. In that case, there was a *minimum* speed while in this case there is a *maximum* speed. What changed in the FBD that gives rise to this important shift?

6.65 The conical pendulum

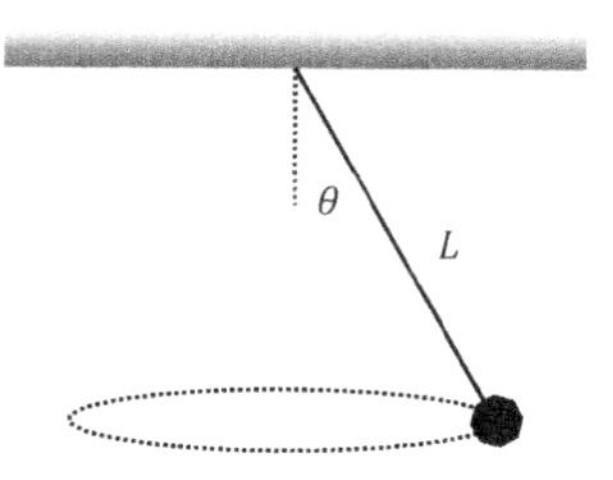

A ball is tied to a string of length L and attached to the ceiling. Let us assume the size of the ball is small compared to the L. The ball is given a push so that is orbits in the horizontal plane with an angle of θ from the vertical as shown in the figure. You may assume the speed is constant.

a) Determine the speed of the ball in terms of L, g, and θ.
b) Suppose the ball was moving faster. What do you expect to happen to the angle?
c) Determine the angle for $L = 1.00$ m & $v = 3.13\frac{\text{m}}{\text{s}}$. Can you determine the angle *algebraically* or are *numerical* methods required?

6.66 Orbiting Nuts…

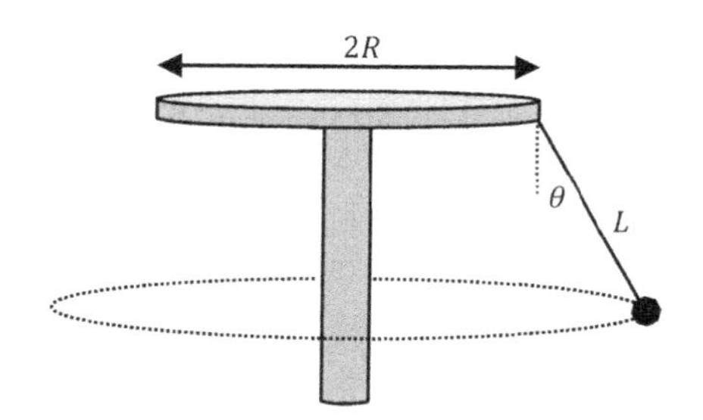

This demo is based on another fair ride. In this ride, people get into swings that are attached to a rotating disk. Initially, when everything is at rest, the swings hang straight down. Gradually, the disk rotates faster and faster. Eventually, at the instant shown in the figure, the ride rotates at a constant rate. As a result the strings form greater and greater angles from the vertical (see figure).

a) Determine an expression for the speed in terms of the radius of the disk and the length of the string.
b) How does the mass affect the angle of the strings? Do amusement parks need to worry about the mass of riders affecting the angle?
c) How is the angle affected if you increase the radius R?
d) Suppose an amusement park ride uses $R = 8.00$ m & $L = 4.00$ m. The ride rotates at 10.0 RPMs. Determine the angle. Can you determine the angle *algebraically* or are *numerical* methods required?

6.67 Water spinner

Suppose you take an aquarium and place it on a turntable. If the system is rotated (with slowly increasing speed) one sees the surface of the water take on a curved shape. The purpose of this problem is to determine an equation for the shape that depends on the rotation rate given by $\omega = \frac{2\pi}{\mathbb{T}}$.

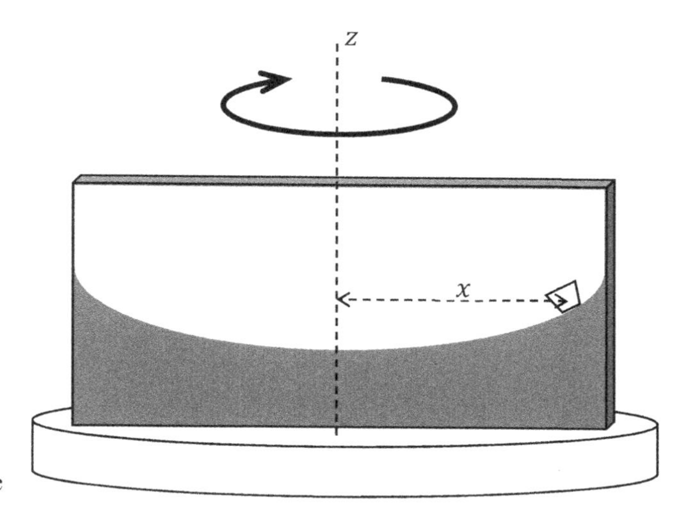

a) Write an expression for centripetal acceleration in terms of ω & r instead of v & r.

b) Suppose a little boat is floating on the surface of the water a distance x from the center of the turntable. Use an FBD of the little boat to determine the force equations for the centripetal and z-directions for the boat.
Hint: the force supporting the boat is called a buoyant force (B) and, in this case, it acts exactly the same way as a normal force would.

c) Show that one finds $\tan\theta = \frac{x\omega^2}{g}$.

d) By definition, this tangent is actually the tangent line to the curve. This is exactly the same thing as saying $\tan\theta = \frac{dz}{dx}$! Using this fact, determine z as a function of x.

e) If you spin the system too fast, this procedure breaks down. Why?

Interesting to note, this device actually has a practical application. Astronomers can make a very large mirror by spinning a cylinder filled with mercury. This type of mirror is used in a zenith telescope. The advantage is that one can make a very high quality mirror in a very large size for a small amount of money. The disadvantages are the toxic fumes and the fact that the mirror can only look straight up!

6.68 A ball swings on a string in a vertical circle. Assume the 1.25 m cord can handle 50.0 N (about 11 lbs.) of force before breaking. The mass on the string is 0.200 kg.

a) At what point in the circular motion is the cord most likely to break?

b) What minimum speed is required to break the string at that point? Note: many times the knot in the string will fail long before the string itself so be careful!

6.69 A ball of mass m swings on a string in a vertical circle. Assume $a_c \gg a_{tan}$ so that the swinging can be modeled as constant speed v.

a) Determine the ratio of the minimum to maximum tension of string during the motion.

b) As $v \to \infty$ does the ratio makes sense?

6.70 Fictitious forces for linear motion (straight line motion)

A cart of mass m_1 is at rest on a horizontal frictionless plane. **Note: m_1 includes the block, the little dude, and all attachments (such as the hanging ball) on it shown in the figure.** It is attached to a hanging mass m_2 over a massless, frictionless pulley as shown in the figure. The entire apparatus is released from rest. Finally, the mass of just the hanging ball is m_3.

a) Determine the acceleration (magnitude) of the system.
b) Once accelerating, the hanging ball will not hang straight down but actually at a small angle (from the vertical). Determine the angle of the hanging ball from the vertical in terms of the masses.
c) Relative to the *earth*, which way does the hanging ball move?

After a very short time, the hanging ball will hang at a fixed angle relative to the little dude. At this point in time the little dude would think the ball is in equilibrium. From the little dude's perspective, it is as if an invisible force is pulling the ball backwards which is balanced by the tension. In reality, the string is pulling the ball forwards and the ball accelerates forwards. We say the invisible force perceived by the little dude is "fictitious".

The fictitious force has magnitude ma. You can actually consider this problem as a Newton's first law problem if you include a force with magnitude ma on the FBD and then set the acceleration in the coordinate system to zero. There are some problems this is actually a really good way to go. One example that comes to mind is a problem we do later on where we think about a car tipping (instead of slipping).

6.71 Balloon in a van

Suppose you are holding a helium balloon in a van. You suddenly step on the gas and accelerate forwards. Consider the acceleration as causing fictitious force of gravity in the x-direction.

a) Do you expect the helium balloon to go forwards or backwards relative to you?
b) What if you slam on the brakes?
c) Why is this different than the ball hanging from a string on an accelerating cart?

6.72 You are sitting in the middle of a car on a slippery bench seat (friction is negligible). The car is travelling quickly on a circular track. From experience you know you would slide to the outside of the turn (left side of the car in the lower figure).

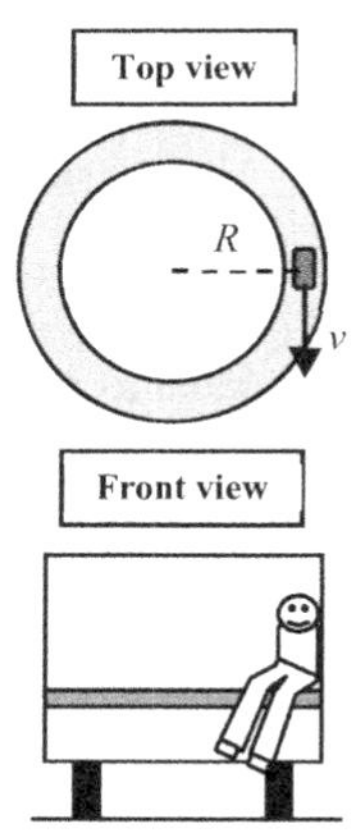

a) Once you are up against the door, what direction(s) does the car push on you?
b) What direction is your acceleration?

6.73 Fishing for Direction

Two peanut butter jars are filled up with water, sealed, and inverted. Inside each jar a fishing float is attached to the lid (at the bottom). The entire apparatus is placed on a turntable and allowed to rotate.

a) When the apparatus is rotated, will the floats go towards the center or towards the edge?
b) What is a practical lab apparatus that operates just like this?

6.74 One time I saw three motorcycles driving around the inside surface of a sphere as part of a show in Vegas. The vertical loop is about 8.00 m in diameter but the center of mass of the motorcycle & rider moves in a circle of diameter 5.00 m. Assume the motorcycle does a perfectly vertical circle while inside the sphere.

a) What minimum speed is required for the motorcycle & rider to execute the circular motion?
b) From an FBD perspective, is this problem more like a ball on a string or going over a hill?
c) **Challenge:** Can I legally get a tax write-off for writing problems like this after attending Vegas shows?

6.75 A car is driven at constant speed over a circular hill and then into a circular valley (with the same radius). To be clear, the car maintains constant speed for the entire track shown above. At the top of the hill the driver experiences a normal force exactly ¼ the size of her weight. How large is the normal force when she reaches the bottom of the valley? Express your answer as a fraction times mg.

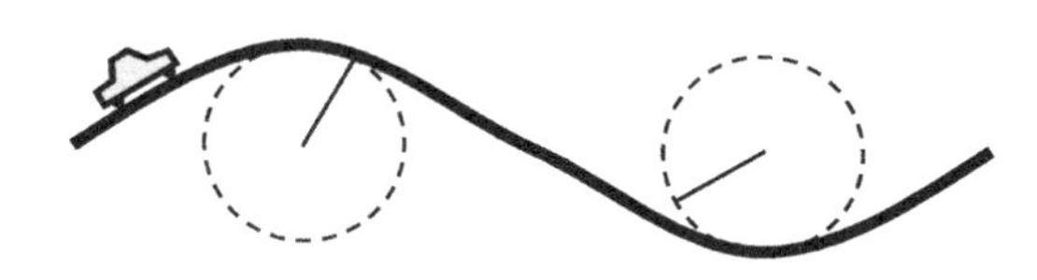

6.76 A ball of mass m is attached to a rotating rod by two strings of length L. The rod rotates at constant rate ω. The strings connect to the rod a distance $h = 1.2L$ apart as shown in the figure; the strings form equal angles with the vertical. A student named Austin notes the ball travels in a circular path in the horizontal plane. The magnitude of the acceleration due to gravity is g. Figure not to scale.

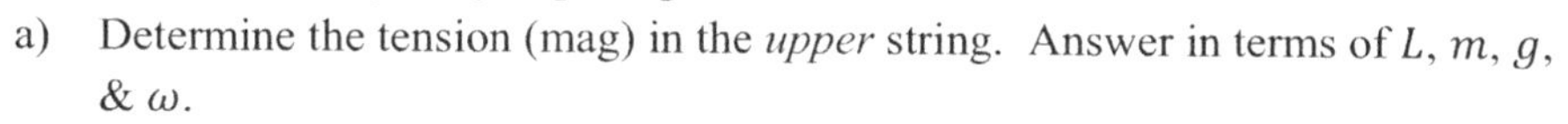

a) Determine the tension (mag) in the *upper* string. Answer in terms of L, m, g, & ω.

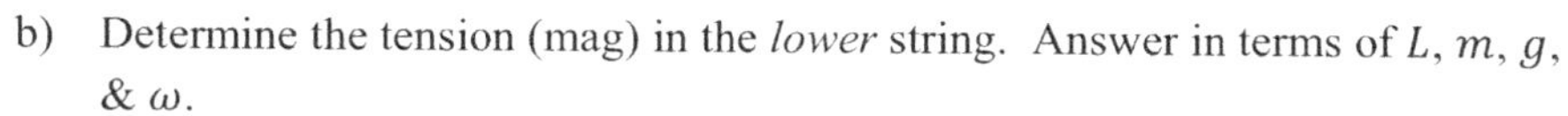

b) Determine the tension (mag) in the *lower* string. Answer in terms of L, m, g, & ω.

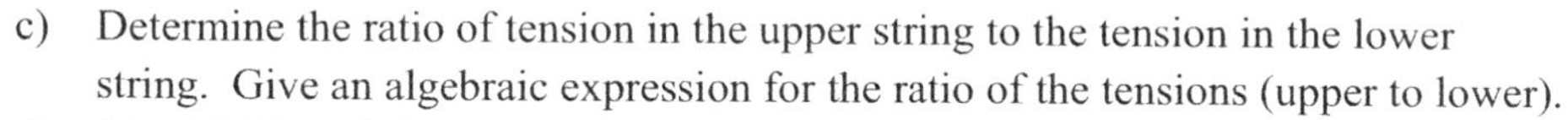

c) Determine the ratio of tension in the upper string to the tension in the lower string. Give an algebraic expression for the ratio of the tensions (upper to lower).
d) Should this ratio be less than, greater than, or equal to 1?
e) Does mass affect the ratio?
f) Does the ratio make sense in the limit of large ω? Hint: consider mg versus ma_c in this limit...

6.77 Challenge: This problem can be found online but the typical solution method is beyond the scope of our course. That said, we can learn about some neat things from this with our basic approach. Come by office hours to discuss if you are curious.

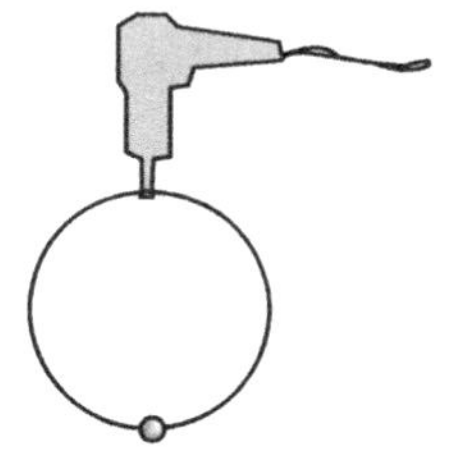

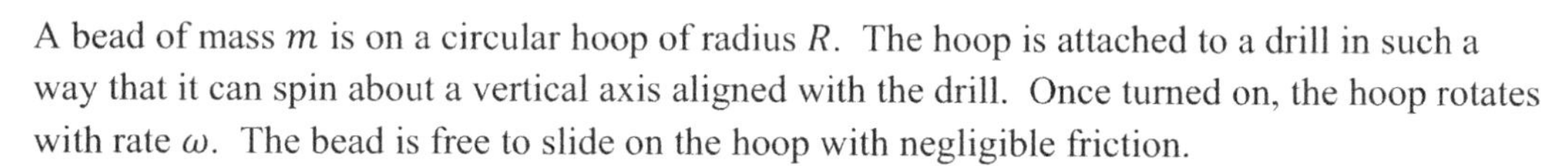

A bead of mass m is on a circular hoop of radius R. The hoop is attached to a drill in such a way that it can spin about a vertical axis aligned with the drill. Once turned on, the hoop rotates with rate ω. The bead is free to slide on the hoop with negligible friction.

a) Assume the drill is turned on and is spinning fairly rapidly. Determine the equilibrium angle of the bead.
b) Now consider a very slow rotation rate with a 6.28 s period for a hoop with radius 10 cm. Do your equations work?

6.78 Two blocks with masses m_1 and m_2 are pushed across a frictionless surface by a constant force $\vec{F}$ angled as shown in the figure.

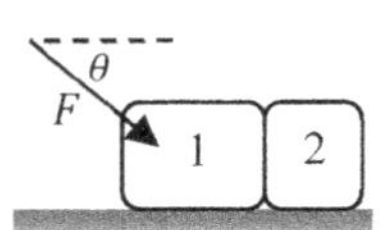

a) Determine the acceleration (magnitude) of the blocks in terms of m_1, m_2, θ, and F.
b) Determine the normal force (magnitude) between the blocks in terms of m_1, m_2, θ, and F.
c) Do your equations make sense for the following cases:
   i. $\theta \to 90°$
   ii. $\theta \to 0°$ and $m_1 \to 0$
   iii. $\theta \to 0°$ and $m_2 \to 0$
d) **Challenge:** Now try redoing this problem if friction is present. No answer in the back for this one, you're on your own. Issues to consider:
   i. For large angles the blocks won't move. This will depend on an interaction of factors.
   ii. For small values of F the blocks won't move.
   iii. You can imagine making a 2D contour plot that would look a lot like a phase diagram from chemistry. There will be large swaths of the plot for which $a = 0$ and other regions where $a > 0$. Figuring out the boundary regions may be non-trivial…I have no clue, haven't thought it through.

6.79 Two blocks with masses m and $2m$ are held in place by the force $\vec{F}$ on the steep incline as shown. To be clear, $\vec{F}$ points perpendicular to the plane. The coefficient of static friction between the two blocks is $\mu_{s12} = 0.5$. The static coefficient between $2m$ and the incline is $\mu_s = 0.2$.

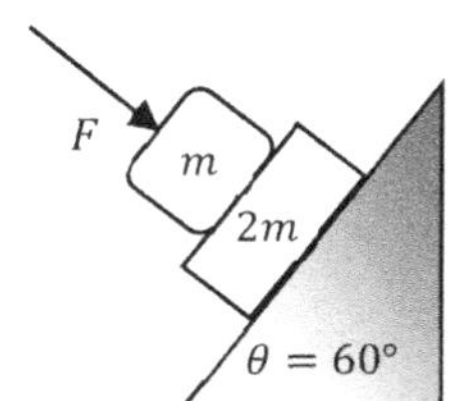

a) Determine the minimum value of F that keeps the system in place.
b) Determine frictional force (magnitude) between $2m$ and the incline. Answer as a decimal times mg.
c) Is $f = \mu_s n$ or is $f < \mu_s n$?
d) Determine the frictional force (magnitude) between m and $2m$. Answer as a decimal times mg.
e) Is $f_{12} = \mu_{s12} n_{12}$ or is $f_{12} < \mu_{s12} n_{12}$?

6.80 A block of mass m is on top of a block of mass $2m$ on an incline. There is negligible friction between $2m$ and the incline. The coefficients of friction between m and $2m$ are $\mu_s = 0.5$ and $\mu_k = 0.3$. A massless, inextensible string connects m to a hanging mass M over a light, frictionless pulley with negligible axle friction.

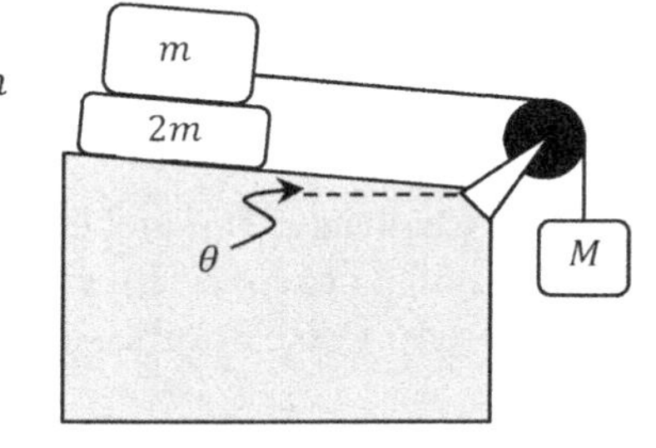

a) Assuming θ is a small angle, determine the largest hanging mass that allows the system to move with $2m$ staying on top of m. Answer as a fraction or decimal with 3 sig figs times m.
b) **Challenge:** if $\theta \geq 61.93°$ will the derivation from part a still work?

Regarding Choices of Coordinate Systems
When doing force problems involving circular motion, you are inevitably faced with thinking about the choice of coordinates and the choice of acceleration direction. The same problem can look very different depending on how one chooses these two things even though, upon proper interpretation, the two give the same results.

Previously, we always used

$$\vec{a} = a_x\hat{\imath} + a_y\hat{\jmath}$$

Whenever possible, we have tried to align our coordinates so that a_x and a_y are always positive. However, as written, either a_x or a_y could be positive or negative numbers. See the following example.

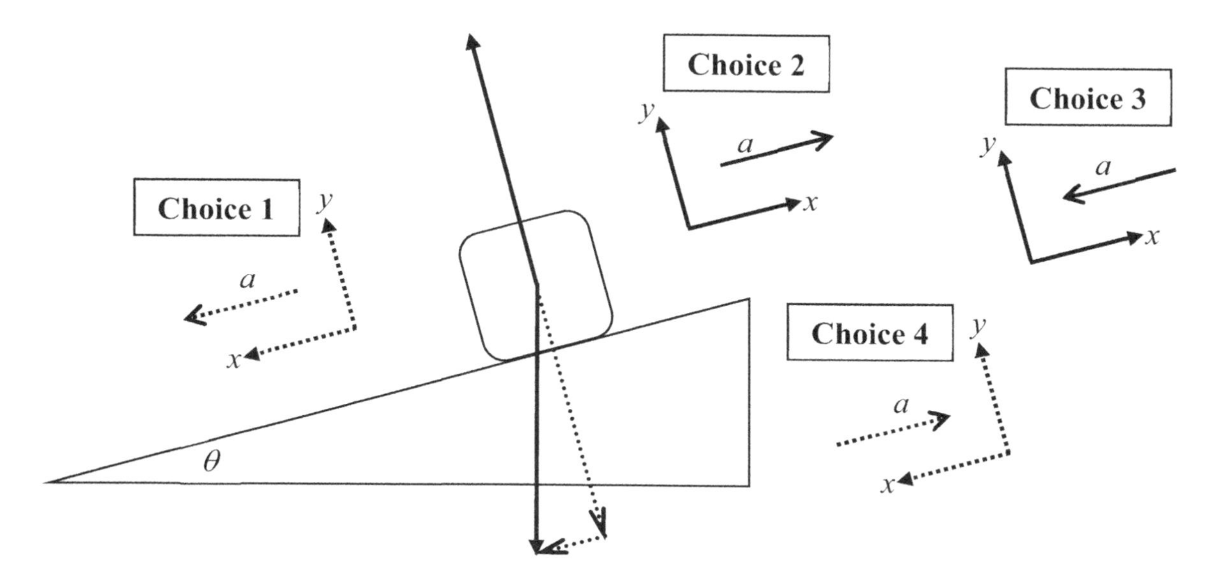

Correctly label the forces in the diagram. Assume friction is negligible. Four choices for the coordinates and acceleration vector are shown. For each choice of coordinates, correctly write down the force equation in the x-direction and solve for a. After doing all four, look at the answers on the next page…

What is the point? There are many ways to choose coordinates and draw acceleration vectors. Depending on your choice, a may be positive or negative. Said another way, it could have an explicit minus sign (one shown in the equation), or an implicit minus (a will be negative when numbers are plugged in).

Notice this: an *implicit* negative value for *a* implies the object, in real life, accelerates <u>opposite the direction drawn</u>.

In later chapters it is often impractical to predetermine the direction of forces (or velocities). So we will make an initial guess on the direction of a force (or velocity) again use the fact that an *implicit* negative value implies <u>opposite the direction drawn</u> in our initial choice.

Answers to previous page. Look carefully at the subtle differences caused by the choices of not only the coordinate system but also the direction of the acceleration arrow.

Choice 1	**Choice 2**	**Choice 3**	**Choice 4**
$mg \sin\theta = ma$	$-mg \sin\theta = ma$	$-mg \sin\theta = m(-a)$	$mg \sin\theta = m(-a)$
$a = g \sin\theta$	$a = -g \sin\theta$	$a = g \sin\theta$	$a = -g \sin\theta$
a is positive, we expect it will point in the direction drawn	a has an explicit negative sign, we expect that in real life the acceleration points opposite the direction drawn	a is positive, we expect it will point in the direction drawn	a has an explicit negative sign, we expect that in real life the acceleration points opposite the direction drawn

If you are wondering when the implicit minus sign appears, one case that jumps to mind is in the case of simple harmonic motion which we will discuss later.

Now consider a ball on a string swinging CCW in a circle that lies in the vertical plane. Assume the ball has mass m and the length of the string is L.

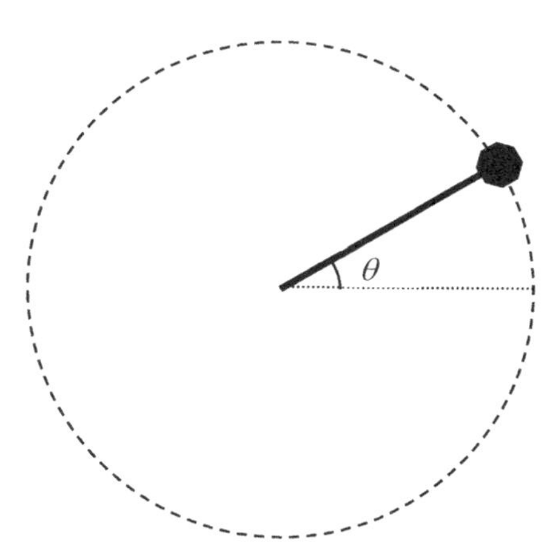

Special Note: The radius of circular motion is from the center of the circle to the center of mass of the ball. We often assume the size of the ball is very small compared to the length of the string. This allows us to assume the radius is L and not $L + R_{ball}$.

In the standard mathematical choice of coordinates the radial direction is outwards and the tangential direction is in the direction of motion of the ball. Notice that in general the acceleration will not have the same angle as the ball!

I have drawn the FBD as well. Notice that, just like the case of the inclined plane, we want to split up the force (or forces) that don't match our coordinates. Notice that the components mg are parallel to the coordinate axes and that the original force mg is the hypotenuse.

Standard Choice of Coordinates

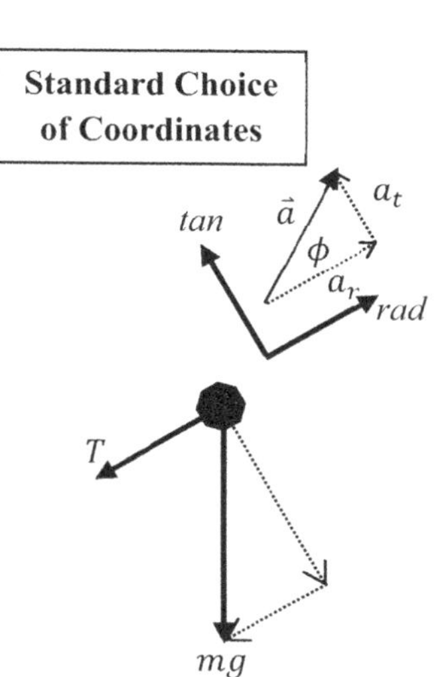

Figure out the upper angle in the mg triangle and correctly label the radial and tangential coordinates of mg.

Determine the force equations and solve for a_t and a_r in terms of m, g, θ, and T. Show that you get

$a_r = -\left(\frac{T}{m} + g\sin\theta\right)$	$a_t = -g\cos\theta$

Now repeat your work for the different choices below. Each case will have slightly different minus signs. Also, watch out as sometimes I used a_r and sometimes a_c! Find both acceleration components for all three choices then check the answers on the next page…

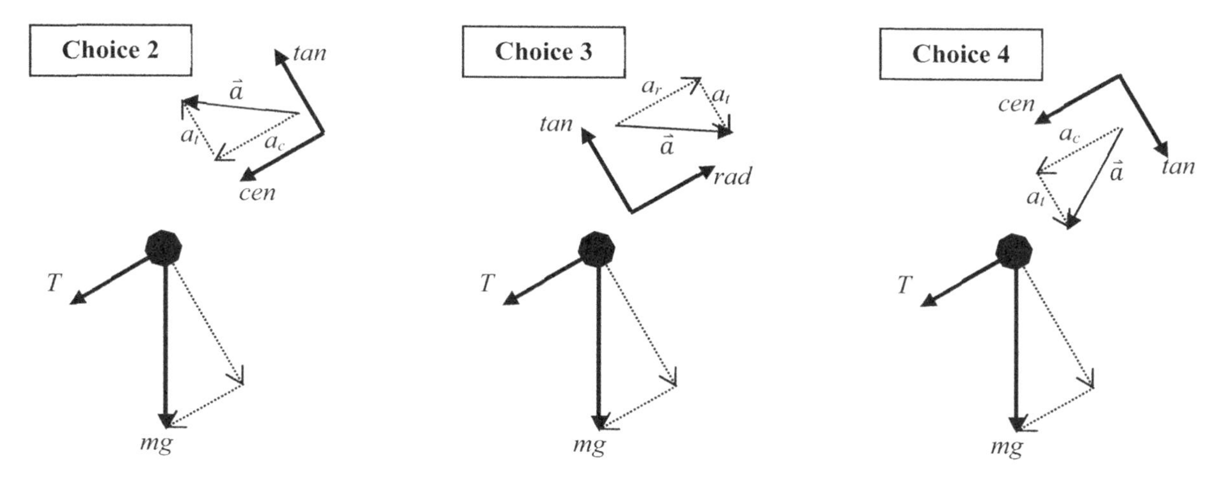

Choice 2	Choice 3	Choice 4
$a_c = \frac{T}{m} + g\sin\theta$	$a_r = -\left(\frac{T}{m} + g\sin\theta\right)$	$a_c = \frac{T}{m} + g\sin\theta$
$a_t = -g\cos\theta$	$a_t = g\cos\theta$	$a_t = g\cos\theta$

What is the point in doing this? Why not choose one method and stick to it? My experience has shown me that mathematicians typically choose the first choice while physicists often choose one of the other choices to reduce minus signs. Some people like their minus signs to be explicit while others like to reduce clutter in the equations and keep the minus sign implicit. Some people like to always draw the acceleration arrows in the direction in the actual direction of acceleration while others are ok with an implicit minus sign and knowing in their head that the accelerations are opposite the direction drawn.

In the end, you should pick a method and stick with it. For this class, I will try to choose the tangential direction in the direction of circular motion. Furthermore, I tend to use a coordinate system that points towards the center rather than radially outwards to reduce a minus sign. This means I usually try to use choice 2. Using this choice $a_c = v^2/r$ and I don't need to think about the minus sign associated with $\hat{r}$.

In addition, the minus sign in a_t makes sense to me because the ball slows down on the way up (when a_t is negative) and speeds up on the way down (when a_t is positive). This works because the ball is on the way up for angles between 270° and 90° and on the way down for angles between 90° and 270°.

Page intentionally left blank.
This gives me space to add content if needed while not having to reformat the cover.

Page intentionally left blank.

Page intentionally left blank.

Page intentionally left blank.

Page intentionally left blank.

Page intentionally left blank.

Equation Sheet

$V_{sphere} = \frac{4}{3}\pi R^3$	$V_{box} = LWH$	$V_{cyl} = \pi R^2 H$	$\rho = \frac{M}{V}$
$A_{sphere} = 4\pi R^2$	$V = (A_{base}) \times (height)$	$A_{circle} = \pi R^2$	$x = \frac{-b \pm \sqrt{b^2 - 4ac}}{2a}$
$C = 2\pi R$	$A_{rect} = LW$	$A_{CylSide} = 2\pi RH$	
160$\underline{9}$ m = 1 mi	12 in = 1 ft	60 s = 1 min	1000 g = 1 kg
2.54 cm = 1 in	1 cc = 1 cm^3 = 1 mL	60 min = 1 hr	100 cm = 1 m
1 cm = 10 mm	1 yard = 3 ft	3600 s = 1 hr	1 km = 1000 m
1 furlong = 220 yards	528$\underline{0}$ ft = 1 mi	24 hrs = 1 day	1 rev = 2π rad = 360°
$g = 9.8\ \frac{m}{s^2}$	$G = 6.67 \times 10^{-11} \frac{N \cdot m^2}{kg^2}$	$P_0 = 1.0 \times 10^5$ Pa	1 eV = 1.60$\underline{2}$ × 10^{-19} J
$1\ N = 1 \frac{kg \cdot m}{s^2}$	$1\ J = 1\ N \cdot m$	$1\ Pa = 1\ \frac{N}{m^2}$	
$x_f = x_i + v_{ix}t + \frac{1}{2}a_x t^2$	$v_{fx}^2 = v_{ix}^2 + 2a_x(\Delta x)$	$v_{fx} = v_{ix} + a_x t$	$r = \sqrt{x^2 + y^2}$
$\vec{A} \cdot \vec{B} = AB\cos\theta_{AB}$	$\lVert\vec{A} \times \vec{B}\rVert = AB\sin\theta_{AB}$	$\sin(A \pm B) = \sin A\cos B \pm \cos A\sin B$	$\cos(A \pm B) = \cos A\cos B \mp \sin A\sin B$
$\vec{v}_{ae} + \vec{v}_{eb} = \vec{v}_{ab}$	$\hat{r} = \cos\theta\,\hat{\imath} + \sin\theta\,\hat{\jmath}$	$\hat{\theta} = -\sin\theta\,\hat{\imath} + \cos\theta\,\hat{\jmath}$	
$a_{tan} = r\alpha$	$a_c = \frac{v^2}{r} = r\omega^2$	$\vec{a} = a_r\hat{r} + a_{tan}\hat{\theta}$	$\vec{a} = a_c(-\hat{r}) + a_{tan}\hat{\theta}$
$\Sigma\vec{F} = m\vec{a}$	$f \le \mu n$	$\vec{F}_G = \frac{GmM}{r^2}(-\hat{r})$	$U_G = -\frac{GmM}{r}$
$TKE = \frac{1}{2}mv^2$	$RKE = \frac{1}{2}I\omega^2$	$U_S = SPE = \frac{1}{2}kx^2$	$U_G = GPE = mgh$
$E_i + W_{\substack{non-con \\ or\ ext}} = E_f$	$\Delta KE = W_{ext.\&\,non-con}$	$W = Fd\cos\theta = F_{\parallel}d$	$W = \int F_x dx$
$\Delta U = -W = -\int_i^f \vec{F} \cdot d\vec{s}$	$F_x = -\frac{d}{dx}U(x)$	$\mathcal{P}_{inst} = \frac{dE}{dt} = \vec{F} \cdot \vec{v}$	$\mathcal{P}_{avg} = \frac{\Delta E}{\Delta t} = \frac{Work}{time}$
$\vec{J} = \Delta\vec{p} = \vec{F}\Delta t$	$\vec{p} = m\vec{v}$	$x_{CM} = \frac{m_1x_1 + m_2x_2}{m_1 + m_2}$	$x_{CM} = \frac{\int x\,dm}{\int dm}$
$\vec{\tau} = \vec{r} \times \vec{F}$	$\Sigma\vec{\tau} = I\vec{\alpha}$	$L = I\omega = mvr_{\perp}$	$\mathcal{P}_{inst} = \vec{\tau} \cdot \vec{\omega}$
$s = r\Delta\theta$	$v = r\omega$	$a_{tan} = r\alpha$	$a_c = \frac{v^2}{r} = r\omega^2$
$I_{\parallel axis} = I_{CM} + md^2$	$I_{zz} = I_{xx} + I_{yy}$	$I = \int r^2 dm$	$\frac{F}{A} = E\frac{\Delta L}{L_0}$
$P = \frac{F}{A}$	$P_{gauge} = P_{abs} - P_{ambient}$	$B = \rho_f V_{disp} g$	$A_1v_1 = A_2v_2$
$P(h) = P_0 + \rho gh$	$P + \frac{1}{2}\rho v^2 + \rho gh = \text{constant}$	$R = \frac{\pi r^4 \Delta P}{8\eta L}$	$F = \eta A\frac{\Delta v_x}{\Delta y}$

Prefix	**Abbreviation**	**$10^?$**		**Prefix**	**Abbreviation**	**$10^?$**
Giga	G	10^9		milli	m	10^{-3}
Mega	M	10^6		micro	μ	10^{-6}
kilo	k	10^3		nano	n	10^{-9}
centi	c	10^{-2}		pico	p	10^{-12}
				femto	f	10^{-15}

Made in the USA
Las Vegas, NV
24 August 2022

53925062R00077